19

II

Second Edition

Edited by
J. T. Greensmith

CONTENTS

ISBN 0-900717-57-2

INTRODUCTION

Cornwall has long been one of the classic areas of British geology on account of the variety of its igneous and metamorphic rocks, and because of the wealth of its mineral deposits. This was one of the first parts of Britain, or indeed of any country, to be studied geologically in detail, and its main features were already known by the early part of the nineteenth century. Nevertheless, many important problems remain to be solved, and there is considerable scope for further investigation.

The distribution of the major rock units is shown in Figure 1.

The largest area in west Cornwall is occupied by Devonian sedimentary rocks which have undergone folding and low-grade metamorphism. Two major stratigraphic divisions are recognised: (1) the Mylor Slate Formation — black slates with fine sandstone bands and occasional pillow lavas, and (2) the Gramscatho Group — an alternation of grey slates and turbidite sandstones.

Unlike the less metamorphosed Devonian sediments elsewhere in Cornwall and Devon, there are no fossils recognisable in the field, and it is only in recent years that micropaleontologists have been able to determine their age from spores and pollen. Upper Devonian

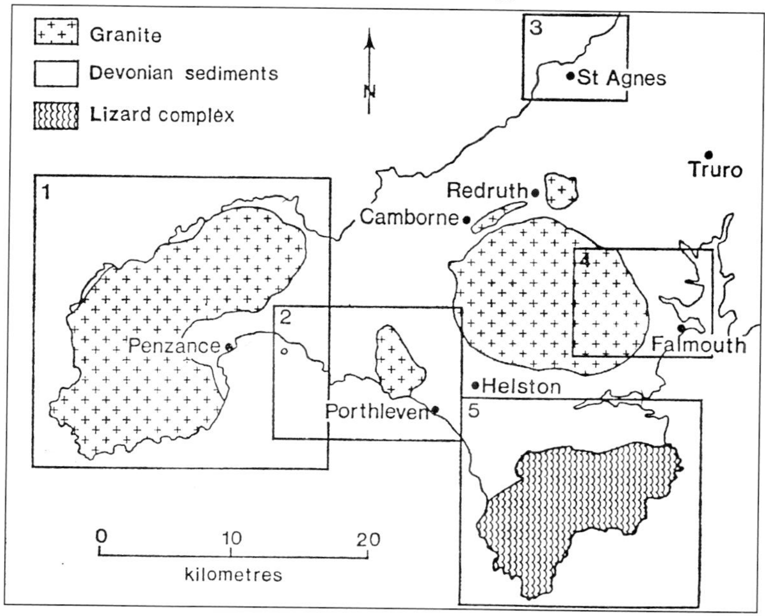

Figure 1: Outline map of West Cornwall, showing the areas covered by the five itineraries.

microfossils have been obtained from both the Mylor slates (Turner, *et al.*, 1979) and the Gramscatho Beds (Le Gall *et al.*, 1985). There is some field evidence that the Mylor slates are younger than the Gramscatho Beds (Leveridge & Holder, 1985), although the relationship between the two facies depends on the interpretation of the structurally very complex section along the south coast between Porthleven and the Lizard. In accounts of Cornish geology written by mining geologists, the term 'killas' is often used to refer to the sedimentary rocks as distinct from intrusions or vein-stones, but more precise petrographic terms such as slate, sandstone, phyllite or hornfels are to be preferred.

The Lizard complex, described in Itinerary V, contains igneous and metamorphic rocks of various kinds, which are nowadays generally interpreted as an ophiolitic assemblage, representing fragments of an ancient ocean floor. The ages of the igneous and metamorphic rocks of the Lizard complex have not yet been fully established, and a full account of the history of the Lizard complex will not be possible until a comprehensive programme of isotopic age determinations has been carried out; the present state of knowledge is reviewed in the introduction to itinerary V.

The granites, which intrude the sedimentary rocks, are of late Carboniferous or early Permian age, and their formation, like the folding of the sediments, is a manifestation of the Variscan (Hercynian) orogeny. The two major intrusions in the area are those of Land's End and Carnmenellis; between them is the smaller intrusion of Tregonning-Godolphin. Granite also outcrops at Cligga Head, St. Agnes Beacon and St. Michael's Mount. Different isotopic methods give slightly different ages, but the results are in the ranges: Land's End 268-275, Tregonning ~280, and Carnmenellis 290-295 million years (Darbyshire & Shepherd, 1985, 1987; Chen *et al.*, 1993; Chesley *et. al.*, 1993). The difference in age between these plutons shows that they were emplaced as separate bodies of magma.

The area is famous for its deposits of tin and copper, which have been worked since prehistoric times. Mining reached its peak in the 1860's, and although it has ceased in other parts of southwest England, it has continued to the present day in the west of Cornwall. The ore deposits are concentrated around the margins of the granites, and the formation of the ore deposits is intimately connected with the intrusion of the latter. At the time of writing, the ore deposits are only being worked at the South Crofty mine near Camborne. China clay has been extensively exploited in this region in the past, but china clay is nowadays only exploited in the St. Austell and Lee Moor districts farther east. Accounts of the old mines are given in Dines (1956) and of their minerals in a beautifully illustrated book by Embrey and Symes (1987).

The youngest deposits of the region are the small patches of Tertiary and Quaternary clay, sand and gravel found at St. Erth, St. Agnes Beacon, Crousa Downs and one or two other places. The Tertiary deposits at St. Agnes are briefly described in Itinerary III. Raised beach, head and blown sand are found at numerous places around the coastline, but the area was never glaciated and the only glacial erratics are a few very large boulders believed to have been deposited by icebergs, of which a well-known example can be seen near Porthleven (Itinerary II).

Many of the conspicuous features of the Cornish landscape can be related to uplift and subsidence of the land and to changes in sea-level during the Tertiary and Quaternary. Notable features are the drowned valleys of the Fal and Helford River estuaries, and the mid-Tertiary erosion surface at 80-120 metres above sea level which is prominent on the Lizard peninsula, to the north of the A30, and west of St. Ives (Walsh *et al.*, 1987).

The Guide is arranged in five sections, covering the areas shown in Figure 1. The localities are described in a geographical sequence, but except for Itineraries III and IV, it would take more than one day to complete each itinerary visiting all the exposures. In addition to the localities described in the Guide, it is worth seeing the collections of the Royal Geological Society of Cornwall (Alverton Street, Penzance) and the Royal Institution of Cornwall (County Museum, River Street, Truro). These collections contain many fine mineral specimens from old mines which are no long accessible.

A general account of the area is given in the British Regional Geology handbook for South-West England (Edmonds, McKeown & Williams, 1969), and the igneous rocks are described in more detail in the Geological Conservation Review volume (Floyd *et al.*, 1993). The 1: 50,000 maps of the Geological Survey which cover this area are 346 (Newquay), 351/358 (Penzance), 352 (Falmouth) and 359 (Lizard), and detailed accounts of the geology are given in the accompanying memoirs. All the localities described in this Guide are shown on the Ordnance Survey 1: 50,000 sheets 203 and 204. Where national grid references are given in the text, the full reference includes the prefix SW.

In planning a visit to west Cornwall it is advisable to consult tide tables before starting out, as most of the best exposures are on the coast. As a general guide, the following table indicates whether it is necessary for the tide to be low to see all the features of a particular locality. A star rating is also given to help in selecting localities for a shorter visit.

Itinerary		*Tide requirement*	*Interest*
I	Land's End	Any	⋆
	Sennen Cove	Low	⋆
	Bostraze	Any	⋆
	Cape Cornwall	Low	⋆⋆⋆
	Botallack	Any	⋆⋆⋆
	Porthmeor	Low	⋆⋆⋆
	Carrick Du	Preferably Low	⋆⋆
	Rosewall Hill	Any	⋆
II	Marazion	Preferably low	⋆
	St Michael's Mount	Low	⋆⋆⋆
	Prah (Praa) Sands	Preferably low	⋆⋆
	Rinsey	Low	⋆⋆⋆
	Tremearne	Low	⋆⋆⋆
	Porthleven	Preferably low	⋆⋆⋆
	Pargadonnel Rocks	Low	⋆⋆
III	Cligga Head	Any	⋆⋆⋆
	Wheal Coates	Low	⋆⋆⋆
	St. Agnes Beacon	Any	⋆
IV	Pendennis Point	Preferably low	⋆⋆
	Swanpool	Low	⋆
V	Jangye Ryn	Low	⋆⋆
	Polurrian Cove	Any	⋆⋆
	Kynance Cove	Low	⋆⋆
	Lizard Point	Low	⋆⋆
	Landewednack	Low	⋆⋆⋆
	Cadgwith	Preferably low	⋆⋆
	Kennack Sands	Low	⋆⋆⋆
	Coverack	Low	⋆⋆⋆
	Porthoustock	Any	⋆⋆
	Porthallow	Low	⋆⋆⋆
	Porthkerris	Any	⋆

ITINERARY I

LAND'S END PENINSULA

The Land's End peninsula corresponds roughly with the outcrop of the Land's End granite intrusion, and the granite does not extend far beyond the present coastline (Figure 2). Patches of the metamorphic aureole are preserved at many places along the north and west coasts, at Longships lighthouse at Land's End, and at Tater-du on the south coast of the peninsula. The hardening of the country rocks by contact metamorphism has evidently made them more resistant to erosion than the unaltered sediments away from the granite.

The Land's End granite provides excellent examples of granite contact relationships, and of the main types of hydrothermal alteration which can

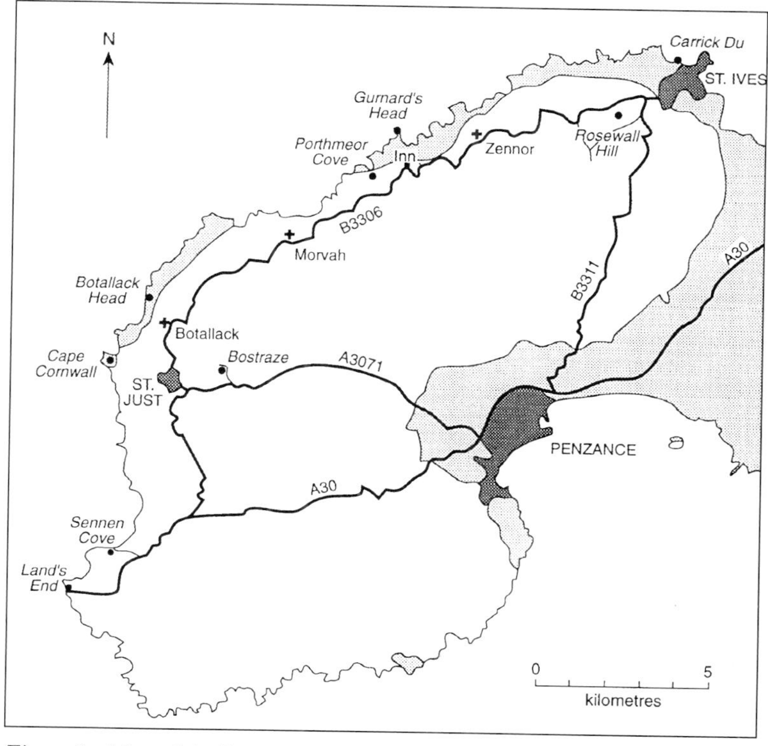

Figure 2: Map of the Land's End peninsula, showing the distribution of granite (white) and country rocks (shaded). Localities described in the text are indicated by dots.

be seen in Cornwall: greisenisation, kaolinisation and tourmalinisation. The country rocks are argillaceous sediments, basic lavas and dolerites which have undergone contact metamorphism and metasomatism.

Land's End.

Land's End is best known to geologists as an example of the influence which jointing has on the scenery. The regular and rectangular joint pattern gives rise to cliffs with a 'castellated' appearance. Despite the apparent uniformity of the rock, there are actually two varieties of granite represented here. On the cliffs adjacent to the State House Hotel, the granite is coarse grained with very large phenocrysts of orthoclase, sometimes exceeding 5 cm in length. A finer granite, with fewer and smaller phenocrysts, forms the cliffs next to the 'First and Last House in England'. The difference between the granites can be seen at a distance by the smoother weathering of the fine variety.

This locality is close to the contact between the Land's End pluton and its country rocks. Longships lighthouse, which is two kilometres directly offshore, actually stands on the country rocks. From here, the contact runs northwards and crosses the prominent headland of Cape Cornwall, which can be seen 6 km to the north. In fine weather, from the top of the cliffs you can also see Wolf Rock lighthouse, which is 15 km to the SW of Land's End. It stands on a small plug of phonolitic lava of lower Cretaceous age, quite unlike any rock exposed on the Cornish mainland.

Sennen Cove.

At Sennen Cove, which is 1.5 km northeast of Land's End, stop at the car park where the road reaches the sea-front and examine the granite exposed on the beach between here and the lifeboat ramp. The granite is a coarse variety rich in K-feldspar megacrysts, which often show a strong alignment. At this locality you can study xenoliths, which are not generally common in the granites of S.W. England. Numerous xenoliths are seen in various stages of digestion by the granite; some show sedimentary banding, some are veined by the granite, and some contain large feldspar crystals similar to those in the granite. Narrow veins of black tourmaline and quartz are common in the granite, and there are a few nodular patches of tourmaline rock surrounded by a lighter feldspar-rich veneer.

Bostraze.

From Sennen Cove, take the road to St. Just (B3306), and just before reaching St. Just turn right on to the main road towards

Penzance (A3071). Stop about 1.5 km along the road when you reach the chimney of Balleswidden Mine on the right-hand side of the road. On both sides of the road there are mine workings and china clay pits. Take the unpaved road on the left, leading to Lower Bostraze china clay pit (ECC International Ltd). This pit is no longer being worked and the lower part is flooded, but it is still possible to examine the kaolinised granite *in situ*.

The china clay workings extend along a NW-SE belt of intense kaolinisation in otherwise fresh granite. The granite was affected by silicification and tourmalinization before being kaolinised, and you may find lumps of hard tourmalinized material left behind when the china clay was extracted.

Cape Cornwall.

Return to St. Just, and at the clock tower in the centre of the village take the road signposted to Cape Cornwall. The road ends at a point overlooking Priest's Cove, which is on the south side of the headland.

In the cove may be seen spotted slates and hornfelses of various types. Look for overturned folds, especially around the boat-launching ramp, quartz veining, and the deformation of hard sandy bands in the more argillaceous slates. Several sheets of microgranite and a few pegmatite veins cut the hornfelses and slates. On the south side of the cove a fault, marked by a small cave, separates the hornfelses from massive granite to the south. The granite is only slightly porphyritic, unlike that of Land's End and Sennen Cove, and is sheared and reddened close to the fault. It is traversed by narrow bands of darkened (greisenised) granite adjacent to thin tourmaline veins. About 100 m south of the contact, the granite contains many irregular bodies of pegmatite up to a metre across, containing tourmaline in large crystals (several centimetres) and radiating masses.

From the end of the metalled road at Priest's Cove walk across a low hill to the beach of Porth Ledden, on the north side of Cape Cornwall (Figure 3). Here there are excellent exposures at low tide of the contact between granite and pelitic hornfelses. The contact dips seawards at varying angles, and veins and sheets of granite penetrate outwards into the hornfels. Near the northern and southern ends of the beach, the granite at the contact is altered to a dark bluish quartz-tourmaline rock, which several metres from the contact passes down into a normal non-porphyritic granite. In the central part of the beach, the contact facies is a coarse granite very rich in potassium-feldspar megacrysts.

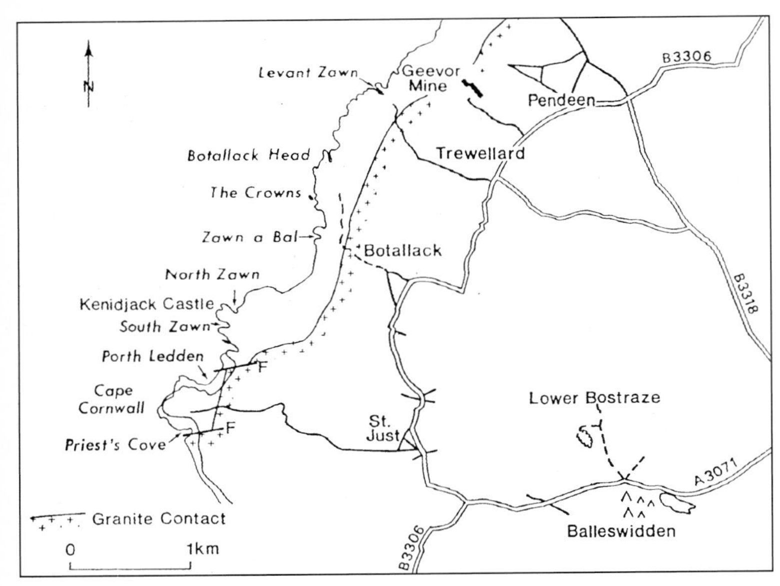

Figure 3: Map of the St. Just area, showing localities mentioned in the text.

Xenoliths can be seen in both the tourmalinised and megacryst-rich granites. In the former, they are represented by irregular patches of black tourmaline-rock. In the later, they include some which contain orthoclase megacrysts similar to those in the enclosing granite. Large dark shadowy patches in the megacryst-rich granite may represent xenolithic material which has been almost completely assimilated. In the megacryst-rich granite, note the conspicuous alignment of the megacrysts and the presence of aplite and quartz-tourmaline veins.

The inclined granite sheets in the hornfels show alternating layers of pegmatite, tourmaline-rich granite and normal granite. They also contain orbicular patches with a narrow dark rim and feldspathic core.

Cross Kenidjack stream at the north end of Porth Ledden beach, follow the stream inland for 100m and climb up the north side of the valley on to the dumps of Boswedden Mine. Notice a trench-like excavation cut into the side of the valley above the dump and continuing on the other side of the valley next to a mine chimney. This is a gunnis, *i.e.* a cavity left by mining a lode up to (or down from) the surface. Follow the gunnis upwards and continue to the top of the hill, from where a well-marked path leads northwards along the cliff top. Pass several small quarries in metabasites (metamorphosed basic igneous rocks), and on reaching the headland Kenidjack Castle climb down the sloping side of South Zawn, facing Cape Cornwall (Figure 3). About half-way down there are good examples of pillow lava, with chert between the pillows, showing that the basic rocks hereabouts are extrusive and not intrusive, as was once supposed. The best exposures are difficult to pinpoint, and a little searching may be necessary. Farther down the cliff there are examples of relatively light-coloured albitised rocks (adinoles). In a small quarry overlooking North Zawn, on the north side of Kenidjack Castle, are the unusual cordierite-anthophyllite-cummingtonite hornfelses described by Tilley (1935).

From here return to the road above Priest's Cove, or continue along the cliffs, crossing the disused rifle range to Botallack. Botallack Mine can be seen in the distance from Kenidjack Castle, its two engine houses standing low down on the face of the cliff.

Botallack.

Approximately 1 km north of St. Just on the main road (B3306), take the left turn which leads into Botallack village, keep left to reach the coast, and at the end of the paved road continue along a track which runs to the edge of the cliffs and forms part of the Cornish coastal footpath. Pass between old mine chimneys on either side of the track and stop after 100m, where there is wall of large granite blocks on the right hand side and the start of a tiny ploughed field on the left hand side. The engine houses of Botallack mine can be seen on the left hand side at the foot of the cliffs (see Front Cover).

There is a deep excavation on the right hand side of the road behind the wall of granite blocks. This is an example of a 'tin floor', *i.e.* a flat-lying tin deposit unlike the more usual steeply dipping lodes. This one is known as Grylls Bunny. There are several tin floors in this area (Figure 4), and they can best be examined 10 m to the left of the road, *i.e.* the closest excavation to the road on the seaward side. They consist of irregular horizontal bodies of tourmaline-rich and cassiterite-rich rock among the metabasites. The cassiterite-rich horizons occur immediately below the tourmaline-rich rock (known to the miners as 'cockle'). The tin floors are overlain by metasomatised basic rocks containing garnet and magnetite.

Follow a path down through highly deformed pelitic hornfelses to the engine houses of Botallack Mine, which are on metabasites. Botallack was one of the best-known mines in Cornwall, having been worked for over a hundred years before it finally closed in 1914. Tin and copper were produced, and a small amount of arsenic. Several lodes were mined, the workings extending for some distance under the

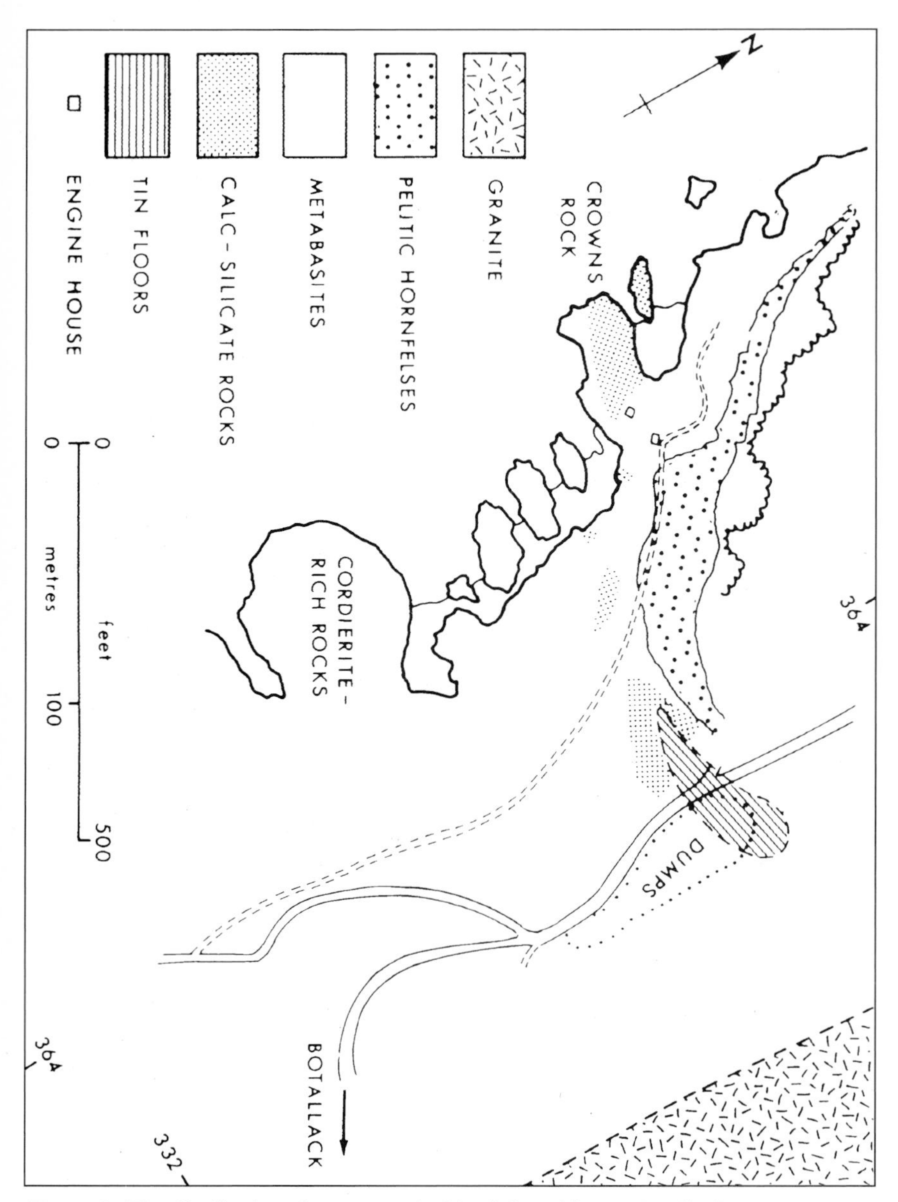

Figure 4: The distribution of metasomatised basic hornfelses at Botallack. From Hall & Jackson (1975).

sea. The two engine houses at this particular spot, shown on the front cover, mark the top of Engine Shaft on Crowns Lode.

The engine houses stand on hornfelsed basic lavas which have undergone considerable metasomatism. Pillow structure can be seen about 50 m north of the upper engine house, adjacent to the cliff path. The pillows show concentric zoning and are separated by light-coloured chert. There are metasomatic bodies of magnetite, garnet, and magnetite-garnet rock around the lower engine house, and specimens of actinolite, epidote, tourmaline and axinite may also be seen here. The garnet rock can be recognised by its pale brown colour, and the magnetite rock by its effect on a compass. Particularly interesting is a dyke-like body of garnet rock which passes underneath the northwest corner of the outer wall of the lower engine house. Some of the metabasic rocks below the lower engine house contain deformed vesicles, showing that they were originally lavas.

In De Narrow Zawn, the inlet which forms the south side of Crowns Rock (on which the lower engine house stands), a belt of iron-staining marks a barren fault (Crowns Guide), filled with quartz-hematite rock. This can be reached by climbing down from the landward side of the lower engine house. An open gunnis can also be seen from Crowns Rock, in the cliffs to the south-southeast.

North of the mine, between Crowns Rock and Botallack Head, pelitic hornfelses are exposed in the upper part of the cliffs, showing a variety of fold-structures. South of the mine, the cliff-top exposures show several types of basic hornfels, including some with large cordierite crystals on the promontory north of Zawn a Bal, from which there is also a good view of the mine.

Two kilometres northeast of Botallack Mine is Geevor tin mine, the last one in the Land's End peninsula to close. There are plans to reopen the surface buildings as a museum of mining.

Porthmeor Cove.

A path leads from the St. Just-St. Ives road (B3306) at Lower Porthmeor (432372) down the valley to Porthmeor Cove. The nearest parking place to this locality is at the Gurnard's Head Hotel (436376).

This spectacular locality shows a great variety of interesting features, especially on the north side of the cove (Figure 5). There is a gently dipping contact between the granite and its country rocks, which are grey, banded, pelitic hornfelses overlying massive dark 'greenstones'. Near the contact, the granite contains numerous parallel tourmaline veins, each bordered by a zone of alteration in which the granite is reddened. There are also a few bands of a different type marked by intense reddening around very narrow quartz veins. Immediately below the contact with the overlying hornfelses, the granite is frequently pegmatitic, and further below the contact is a layer of dark quartz-tourmaline rock a few centimetres thick.

The hornfelses are hard and splintery. The pelitic variety sometimes contains dark spots, and the basic variety sometimes contains coarsely crystalline actinolite. Many veins and sheets of granite, microgranite and pegmatite penetrate from the granite into the hornfelses. The larger

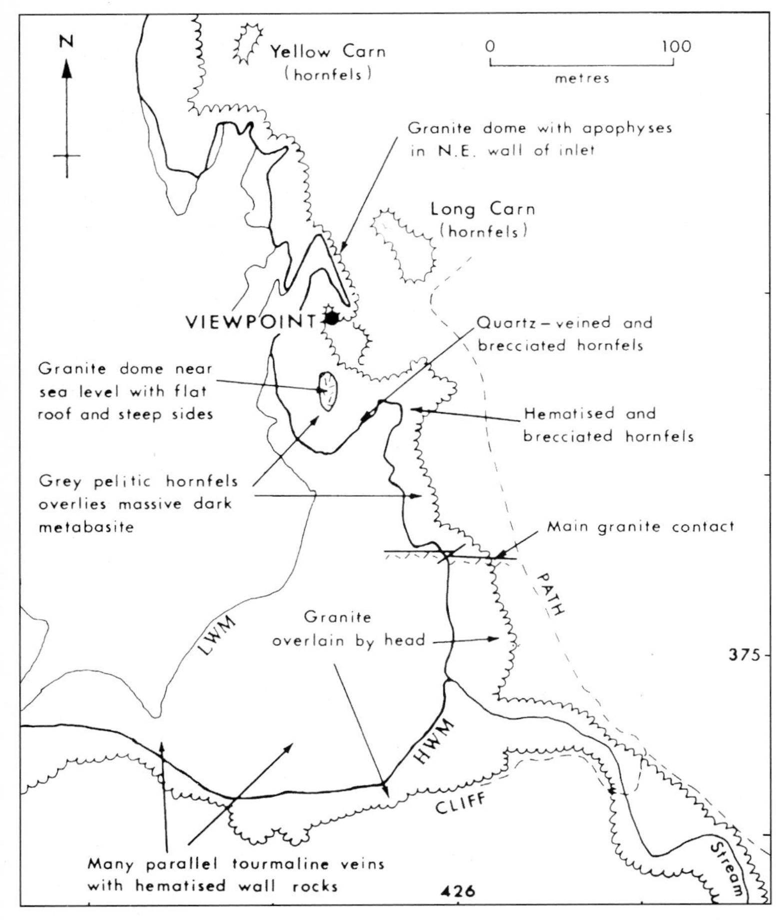

Figure 5: Details of the exposures in Porthmeor cove. From Hall & Jackson (1975).

sheets contain xenoliths, pegmatitic segregations and clots of tourmaline. Most of the granite sheets are fine grained, but the larger ones contain feldspar phenocrysts similar to those in the main body of granite.

A gully in the north side of the cove marks a zone of brecciated, quartz-veined and reddened hornfels. Beyond this zone, granite veins are very numerous. One large mass of granite has the form of a small dome tens of metres across, with a flat roof and steep sides; for about a metre below the roof the granite contains horizontal bands of pegmatite and aplite, which pass down into homogeneous granite. From the view-point shown in Figure 5, another dome is visible in the vertical cliff to the northeast. This exposure shows apophyses from the granite extending both vertically and horizontally into the overlying and adjacent hornfelses, and you can see large angular blocks of hornfels which have sunk into the granite from its roof.

After returning to the main road, continue your journey northeastwards towards St. Ives. Between Porthmeor and St. Ives, the view on the left hand side of the road displays a superb example of an ancient erosion surface about 110 metres above the present sea-level. The boundary between this cultivated flat ground and the rocky hillside to the right of the road has been described as a former coastline backed by degraded cliffs, but Walsh *et al.* (1987) argued that the erosion surface was subaerial and not marine. This geomorphological feature (the change of slope) bears no relation to the boundary between the granite and its country rocks, which is near to the present coast line.

Carrick Du.

Continue into St. Ives, and take a left turn signposted to Porthmeor Beach, where there is a car park. St. Ives is very congested during the summer months, and severe traffic restrictions apply, so you may have to take advantage of the park-and-ride scheme.

Carrick Du is the promontory at the western end of Porthmeor Beach. The fine-grained, black rocks seen on the beach are basic hornfelses, *i.e.* basic igneous rocks that have been metamorphosed. The rocks seen on the foreshore at the western end of the beach are fine-grained and black with no obvious structure, but at Carrick Du headland they are seen to be pillow lavas. This is much easier to see on the west-facing sides of the various small promontories that make up the headland. A satisfactory view is obtained from the coastal footpath. You can get closer by climbing down the west side of the promontory nearest to the beach, but this is only safe at low tide and when the sea is calm. Despite the relatively undeformed nature of the pillows, these rocks have undergone both regional and contact metamorphism and can be classified as albite-actinolite hornfelses (Floyd *et al.*, 1993).

Rosewall Hill.

Leave St. Ives in the direction of St. Just on the B3306 road, turn left onto the Penzance road (B3311), and after 100 m turn right into Towednack Road. About 1 km along this road, several footpaths lead up to the top of Rosewall Hill, which is on the right. A line of surface workings, shafts and mine dumps of Rosewall Hill mine extends up the hill in an ENE-WSW direction. Taking care not to fall into any open workings, ascend the hill to the lower of the two chimneys near the top, where a deep cavern has been left by the removal of ore down from the surface. Several dumps surround this excavation. In addition to coarse porphyritic granite, which is the country rock, there are good examples of tourmalinised and chloritised granites, and of cassiterite-bearing quartz-tourmaline rock, which is the actual lode material.

Climb to the top of the hill, where the widely spaced jointing has enabled the granite to weather into typical tors. The alignment of the large phenocrysts in the granite is very noticeable. From the top of the hill there is a good view across the 110 metre erosion surface towards the coast. One kilometre to the northwest is another hill, also made of porphyritic granite, on the seaward side of which is the large cluster of buildings known as Trevalgan Farm (490401), after which trevalganite is named. This is a dark rock consisting of large orthoclase phenocrysts and smaller quartz crystals in a fine-grained groundmass of black tourmaline; blocks of this material have been used in constructing the walls of the farm buildings. It is very similar to the tourmalinised granite in Bostraze china clay pit and is presumably a product of local hydrothermal alteration of the granite.

The mineralisation of Rosewall Hill conforms to the pattern of zoning characteristic of an emanative centre. Main Lode of Rosewall Hill Mine continues to the northeast as Standard Lode of St. Ives Consols (around the junction of the B3306 and B3311 roads) and as Main Lode of Trenwith Mine (in St. Ives town). Within the granite (at Rosewall Hill) the lode mainly carries tin mineralisation; near the granite contact (at St. Ives Consols) the mineralisation is of tin and copper; in the country rocks beyond the contact (at Trenwith) the mineralisation is of copper (sulphides) and uranium (pitchblende) with traces of lead (galena) and zinc (sphalerite), but with relatively little tin. This sequence of minerals is said to indicate deposition from an ore-forming fluid whose temperature decreased away from the granite contact.

ITINERARY II

MOUNT'S BAY

The coast between the Land's End and Lizard peninsulas offers a complete section through the Tregonning-Godolphin granite (Figure 6) and exposures of several smaller igneous intrusions. The itinerary begins at Marazion and ends at Porthleven.

Marazion.

The principal purpose of starting at the village of Marazion is to visit the offshore island of St. Michael's Mount, but the foreshore of Marazion itself is of interest and may be inspected while waiting to cross to the island, which is only accessible at certain times of the day.

Between the causeway to the island and the small harbour of Marazion the shore is composed of slate or phyllite, into which are intruded several bodies of dolerite, some concordant, some discordant, and some drawn into lenses by movement of the softer slates. From the end of the harbour breakwater notice the vertical cliffs of Quaternary head deposit extending along the coast to the east.

St. Michael's Mount.

At St. Michael's Mount, a small area of granite is exposed in contact with slate, representing the northern tip of an intrusion of uncertain size, the remainder of which is concealed under Mount's Bay. The granite is of interest because it contains a great abundance of greisen-bordered veins, and is rather similar to the Cligga Head granite described in Itinerary III. The island forms part of the St. Aubyn estates, now administered by the National Trust, and is the home of the St. Levan family. It is therefore necessary to obtain written permission before visiting the island, by writing to the General Manager (St. Michael's Mount), Manor Office, Marazion, Cornwall or by visiting the office. Permission will only be granted on the condition that hammers are not carried. The island is accessible on foot by a causeway which is passable for several hours a day at low tide, but on Wednesdays and Fridays throughout the year a ferry operates during the hours of high tide.

Walk around the shoreline of the island starting on the east side. Here you can see the contact between the slates and the granite, the latter forming the southern half of the island. Granite veins penetrate into the slate, which has been altered by contact metamorphism into a mica-schist. A swarm of roughly parallel east-west quartz veins cuts the granite, which is altered to greisen on either side. Some of the greisen

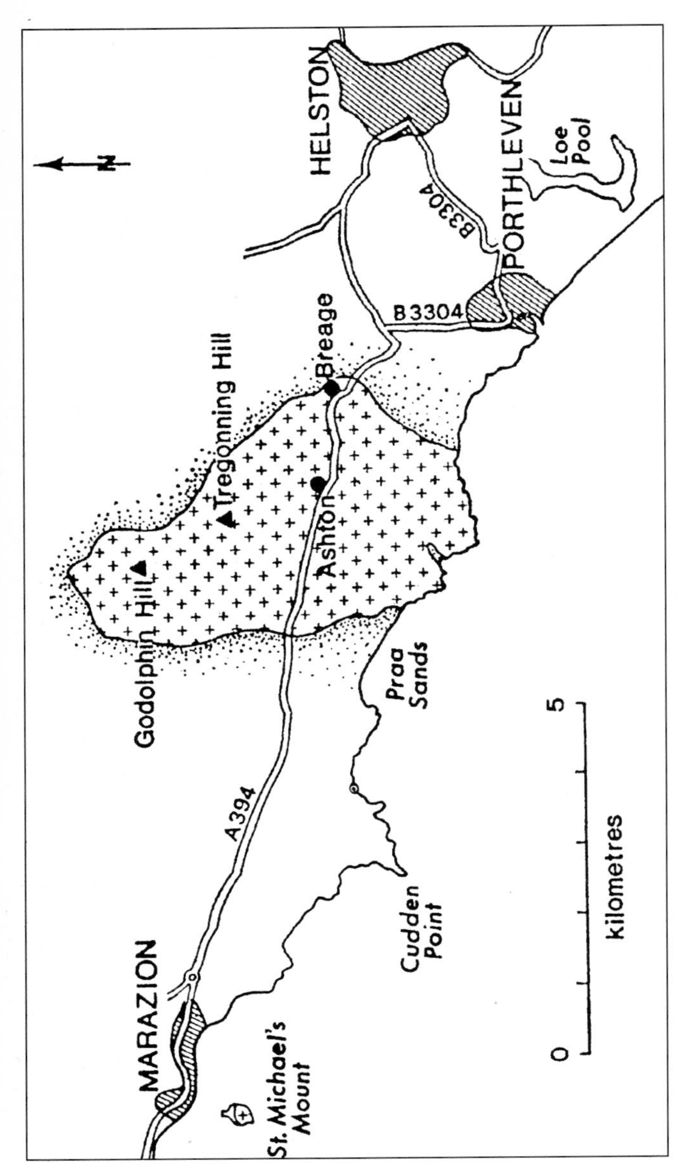

Figure 6: Map of the coast of Mounts Bay between Marazion and Porthleven, showing the outcrop of the Tregonning-Godolphin granite and its metamorphic aureole.

bands are up to a metre in width and contain several quartz veins. They sometimes diverge and coalesce, occasionally spreading out into a patch of massive greisen. The veins are mineralised and black platy wolframite and green oxidation minerals of copper can be observed. Locally a greenish coloration shows the copper minerals to be disseminated in the greisen as well as concentrated in the veins. In places the quartz and greisen are more resistant to weathering than the granite, and stand out slightly from the surface. The granite is most readily weathered immediately next to the greisen bands, and the granite-greisen junction is sometimes weathered into a groove.

On the west side of the island, two sets of greisen-bordered veins can be seen, intersecting at angles between 30° and 45°. Aplite dykes can also be seen here. The contact between granite and country rocks is again seen, and granite veins penetrate into the sediments. Some of the granite veins contain angular xenoliths of altered sediment. Greisen-bordered quartz veins are found even in the granite that penetrates the slate, but although the latter contains veins of quartz there is no indication of alteration at their margins. It may be noticed that quartz veining in the slates took place at several different times, as some of the veins are cut off by the granite while others penetrate into the granite. Additional information on this locality is given by Hosking (1957).

Prah (Praa) Sands.

Six kilometres east of Marazion, take the turning off the A394 signposted to Prah (Praa) Sands, where there is a car park adjacent to the beach. At the western end of the beach the first rock exposed is a quartz-porphyry. This is a dyke which extends for several kilometres inland towards the Land's End granite, and is a good example of the quartz-porphyry dykes which occur throughout Cornwall in association with the granites, and are locally called 'elvans'. The rock contains large phenocrysts of orthoclase and quartz in a very fine-grained groundmass of the same minerals plus muscovite.

This rock has been studied by Stone (1968), who showed that unlike the granites it does not have a composition near the 'minimum' melting composition for acid igneous rocks, but instead has an exceptionally high K/Na ratio. This is a feature of many Cornish quartz-porphyry dykes and has been attributed to ion exchange with a hydrothermal fluid, although in the field the rock gives no appearance of having suffered hydrothermal alteration.

Beyond the dyke, which has chilled margins at its contacts, are highly contorted slates containing numerous deformed quartz veins. Two varieties of slate can be distinguished, one soft, flaky and grey in

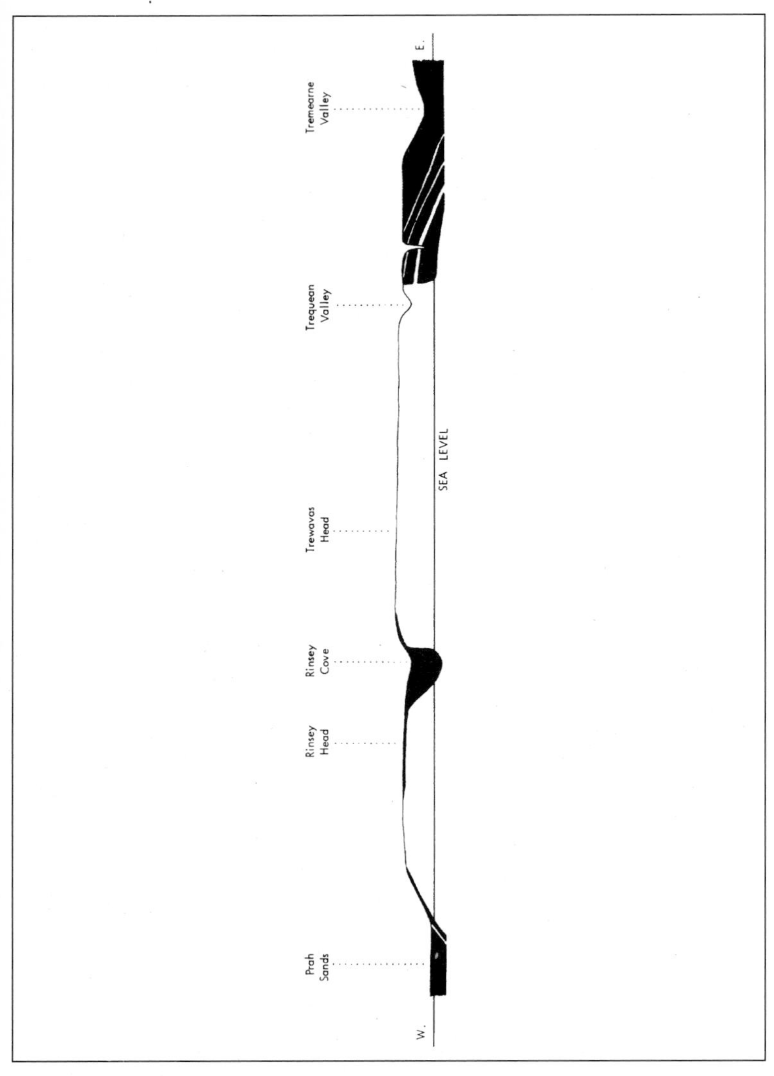

Figure 7: A profile of the coast between Praa Sands and Tremearne, showing the form of the Tregonning granite. The top of the cliffs at Rinsey Head follows the contact between granite (white) and country rocks (black). Granite sheets extend into the country rocks from the eastern contact. The length of the section is 4 kilometres; the maximum height of the cliffs is 90 metres.

colour, the other hard and black. Although it is not visibly spotted, the hardness of this slate is due to contact metamorphism by the Tregonning granite. The aureole of the granite is unusually wide (wider than shown in Figure 6), and signs of metamorphism extend for about 3 km to the west and north of the intrusion, suggesting that it underlies a much wider area than its surface outcrop.

About 20 metres west of the beach cafe there is a section through Quaternary deposits, showing 30 cm of blown sand, overlying 20 cm of peat, overlying 2 metres of head. The head deposit was formed by the movement of surface material under periglacial conditions, and contains many fragments of vein quartz together with slate and elvan and is crudely stratified. The top 50 cm of head contain many rootlets and is altered into a grey clay, which may be regarded as a seat-earth below the peat. Taylor and Goode (1987) obtained a radiocarbon date of 1805 ± 100 years B.P. for a sample of peat from Praa Sands, and suggested that the peat accumulated in reed beds ponded behind the advancing sand dunes that eventually buried it. The head deposit extends the whole length of the beach, covering the underlying slate and granite and forming a low cliff.

At the eastern end of Prah Sands, the Tregonning-Godolphin granite is exposed, and a continuous section extends along the coast for 3 km (Figure 7). This granite shows many interesting features and has been studied in detail by Stone (1975).

At the eastern end of Prah Sands, the contact between granite and slates is exposed at low tide. The contact dips at an angle of about 30° to the horizontal, and beneath its roof the granite displays a remarkable banding in which granite, pegmatite and aplite alternate in layers a few centimetres thick, all parallel to the roof. A few xenoliths of tourmalinised hornfels can be seen. Keep to the shoreline as far as the tide permits and then climb to the cliff path, which leads round the headland into Rinsey Cove. Some of the granite exposures seen along the way are relatively homogeneous, while others show banding indicating the proximity of the roof.

Rinsey Cove.

Rinsey Cove can be reached along the coastal footpath from Praa Sands or by road via the small hamlet of Rinsey (594273), where it is usually possible to park a coach. Cars can be parked a little closer to the coast at the point indicated in Figure 8.

Figure 9 shows details of the exposures in Rinsey Cove. The cove was formed by the erosion of a roof pendant which descends to sea-level at this point. Overlooking the cove is an old mine, Wheal Prosper, whose dumps contain granite, quartz veins and tourmaline-bearing veinstone. From here a path leads down to the sea.

On the west side of the cove, hornfelsed slates overlie the granite with a sharp contact dipping eastwards at 30°. Banding can be seen in the granite below the contact, and there are also irregular pockets of pegmatite rich in acicular black tourmaline. There are numerous tourmalinised xenoliths, some with sharp margins and other diffuse. Many very diffuse dark patches near the roof of the granite may also be

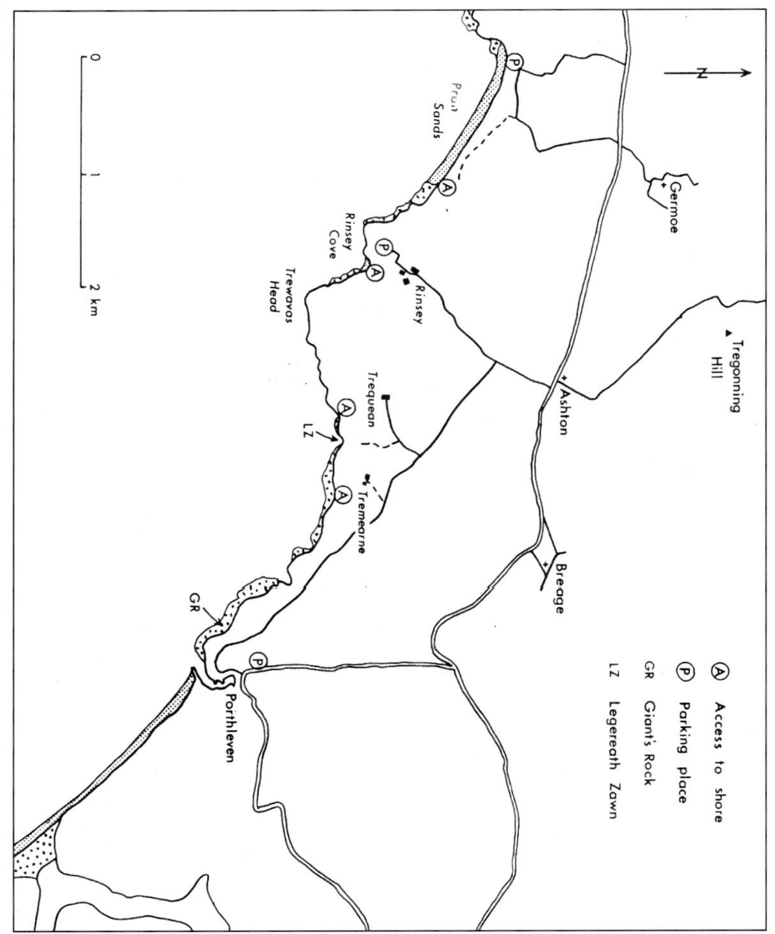

Figure 8: Map of the coast section of the Tregonning granite, showing localities mentioned in the text.

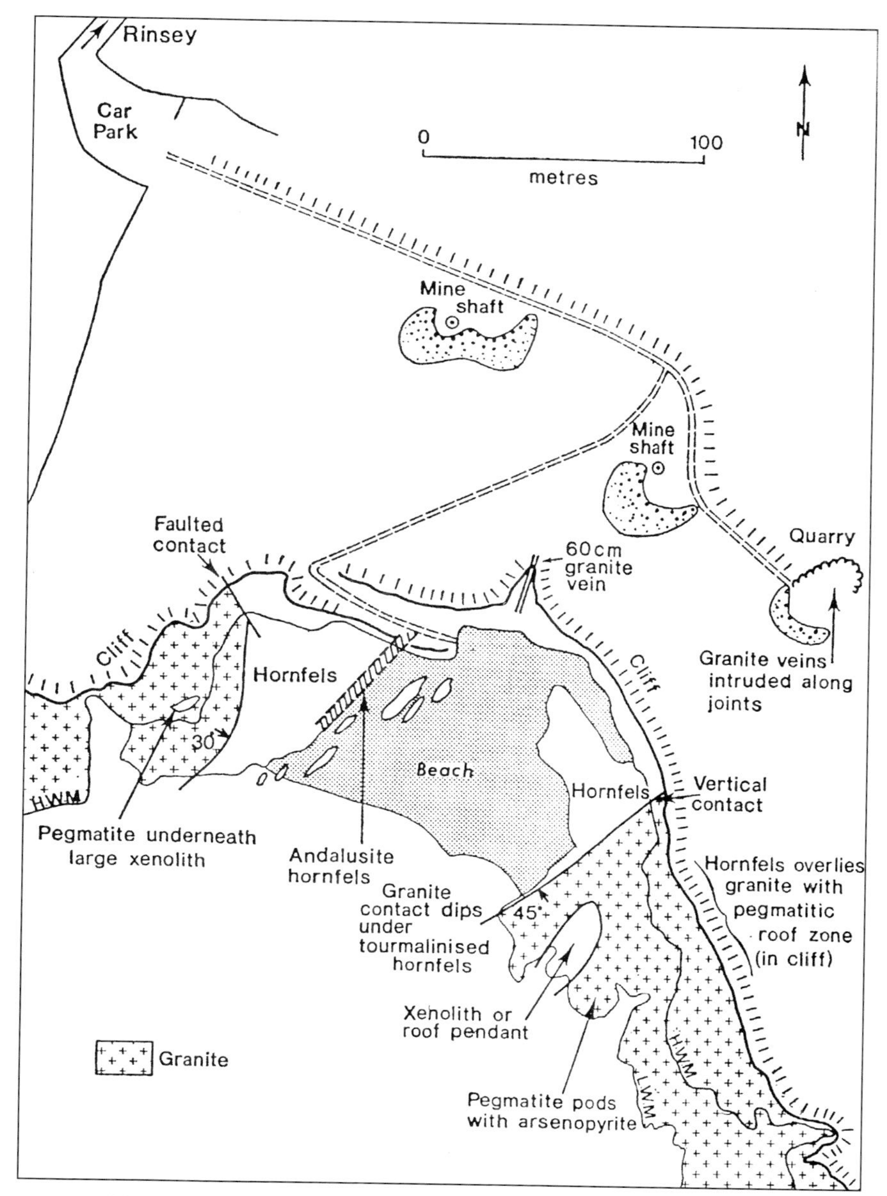

Figure 9: Map showing details of the exposures in Rinsey Cove (from an original map by N. Jackson).

due to the assimilation of xenolithic material. Several sheets of pegmatite and numerous quartz veins of more than one generation cut the hornfelses. Some of the hornfelses are rich in andalusite.

On the east side of the cove, the contact is nearly vertical and granite extends upwards nearly to the top of the cliffs, although a roof of slate is still present. A dark border about 2 cm thick shows that the slates are intensely altered immediately next to the contact. On the seaward edge of the wave-cut platform there is a small roof pendant of slate in the granite, and the granite nearby contains pockets of pegmatite with large crystals of arsenopyrite. Near the top of the cliffs on the eastern side of the cove, the roof of the granite is almost horizontal, and long black crystals of tourmaline project down into the pegmatitic roof facies from the overlying tourmalinised slate.

On the east side of Trewavas Head, 700 m southeast of Rinsey Cove, are the ruined engine houses of Trewavas copper mine. From the seaward side of the first engine house, a distant view is obtained of Tremearne cliffs, where several very thick horizontal granite sheets, light in colour, can be seen cutting the slate at the eastern contact of the granite.

Tremearne.

The coastal section across the eastern contact of the Tregonning granite shows very interesting features but is difficult of access. The main contact reaches the sea at Trequean cliff, which is at the end of the valley below Trequean Farm. From the contact, several thick granitic sheets run into the slates and are exposed in the face of Tremearne cliff, south of Legereath and Tremearne. Access to the shore can be obtained at two points, shown in Figure 8.

A steep gully descends from the cliff path 50 m east of the lowest part of Trequean Farm valley. The descent is rather dangerous and the shore section is overhung by a head deposit containing loose boulders. The exposure shows a nearly vertical contact between the granite and hornfelsed slate. From the wave-cut platform there is a very good view of the granitic sheet of Tremearne cliff, but it is not normally possible, even at low tide, to cross the stretch of water (Legereath Zawn) at the base of the deep gully below Legereath Farm.

Easier access to the shore is obtained at the valley leading down from Tremearne Farm, where there are steps down from the coastal path to the beach. The country rocks here are hard spotted slates. From here, walk westwards across the foreshore (Megiliggar Rocks) to examine the eastern end of one of the granitic sheets. These sheets consist of layers of pegmatite and aplite, with numerous xenoliths. This is probably the best locality in England for seeing granitic pegmatites and has been described in detail by Stone (1969).

Porthleven Sands.

At Porthleven there is a fine coast-section to the east of the harbour. From the harbour entrance there is a good view of the low peneplain of the Lizard towards the southeast. As well as examining the cliffs, notice the beach of Porthleven, which is composed of small chert pebbles. Reid suggested in 1904 that these pebbles might be derived from a possible Eocene outcrop beneath Mount's Bay. A subsequent submarine survey showed that Cretaceous and Eocene deposits do in fact outcrop on the sea-floor about 50 km to the south.

The exposures in the cliffs show the Mylor beds, which are dark grey or blue slates with many thin lighter sandier bands repeated every few centimetres. This banding or striping is characteristic of the Mylor beds and is best developed below the coastguard lookout (500 m east of the harbour wall), where millimetre-scale layering can also be seen in the sandy bands. There are numerous quartz veins in the slates, some of which contain small amounts of sulphide minerals.

The slates are cut by several dolerite sills. These are massive greenish rocks, or brown when weathered, and they are largely recrystallised to 'epidiorites', consisting of actinolite, albite, chlorite, calcite and other alteration products. Examples of spheroidal weathering can be seen. The largest dolerite sill, just east of the coastguard lookout, is about 20 m thick and is relatively coarse-grained with crystals up to a centimetre in length.

Between about 500 m and 1000 m east of the harbour entrance there are several adits in the cliffs. These are part of the workings of Wheal Penrose, one of the few mines in Cornwall to have been worked mainly for lead, and believed to have been exploited since Roman times. The lode is a 'cross-course' *i.e.* it has a NNW – SSE trend, at right-angles to the predominant lode direction in Cornwall, and like other cross-courses it carries relatively low temperature lead-zinc mineralisation.

From a small beach backed by a patch of head deposit ascend by steps to the road above from where you can return to the centre of Porthleven past the dumps of Wheal Penrose, where specimens of siderite, galena and sphalerite may be collected. Alternatively you can continue along the coast to see Loe Pool, formed by the blocking of the Cober estuary by a huge shingle bank.

Pargodonnel Rocks.

From Porthleven Harbour walk up the road which ascends the hill on the west side of the harbour until you reach the cliff-top path. Walk along the path until you see a wayside cross about 100 m ahead. The shore below you on the left is a wide wave-cut platform, exposed at low tide, called Pargodonnel Rocks. You will notice a very large boulder resting on the platform. Climb down to the shore by the path and steps from the cross and walk towards the boulder, which is known as Giant's Rock.

Near the bottom of the steps, around the entrance to a small cave, are patches of iron-stained, cemented sand and gravel adhering to the slates. These are the remains of a raised beach deposit. Elsewhere in this area such raised beach deposits are sometimes seen to be overlain by head, indicating a considerable age, since the latter presumably dates from at least the late-glacial period. On the shore there are numerous small boulders including slate, granite and vein quartz, which have possibly been washed out of the former raised beach.

Giant's Rock is quite exceptional. It is about 3 m in length, weighs about 20 tons, and is lodged in a rock pool from which it is not moved by even the heaviest storms. It is highly polished, brown in colour, and is composed of garnetiferous gneiss which cannot be matched with any other rock in Britain. It is presumed to have been stranded by an iceberg during the glacial period. The gneissic banding and garnet crystals (up to a centimetre across) should be examined **without hammering.**

ITINERARY III

ST. AGNES DISTRICT

The St. Agnes district (Figure 10) has been an important mining area in the past. It includes two small granite bosses, those of Cligga Head and St. Agnes Beacon, and their metamorphic aureoles. The mineralisation is concentrated around the two intrusions, which are therefore regarded as 'emanative centres'. The area is also notable for the only outcrop of Miocene sedimentary rocks in Britain.

Cligga Head.

The excursion starts from Perranporth. Leave Perranporth by the St. Agnes road (B3285). The road ascends to a plateau at about 100 m. Before reaching the top of the hill, look for the entrance on the right to Cligga Head airfield, now used by the Cornish Gliding Club. Park your vehicle, if any, before reaching the runway, which may be in use. Walk across the northern end of the airfield to the remains of Cligga Head Mine.

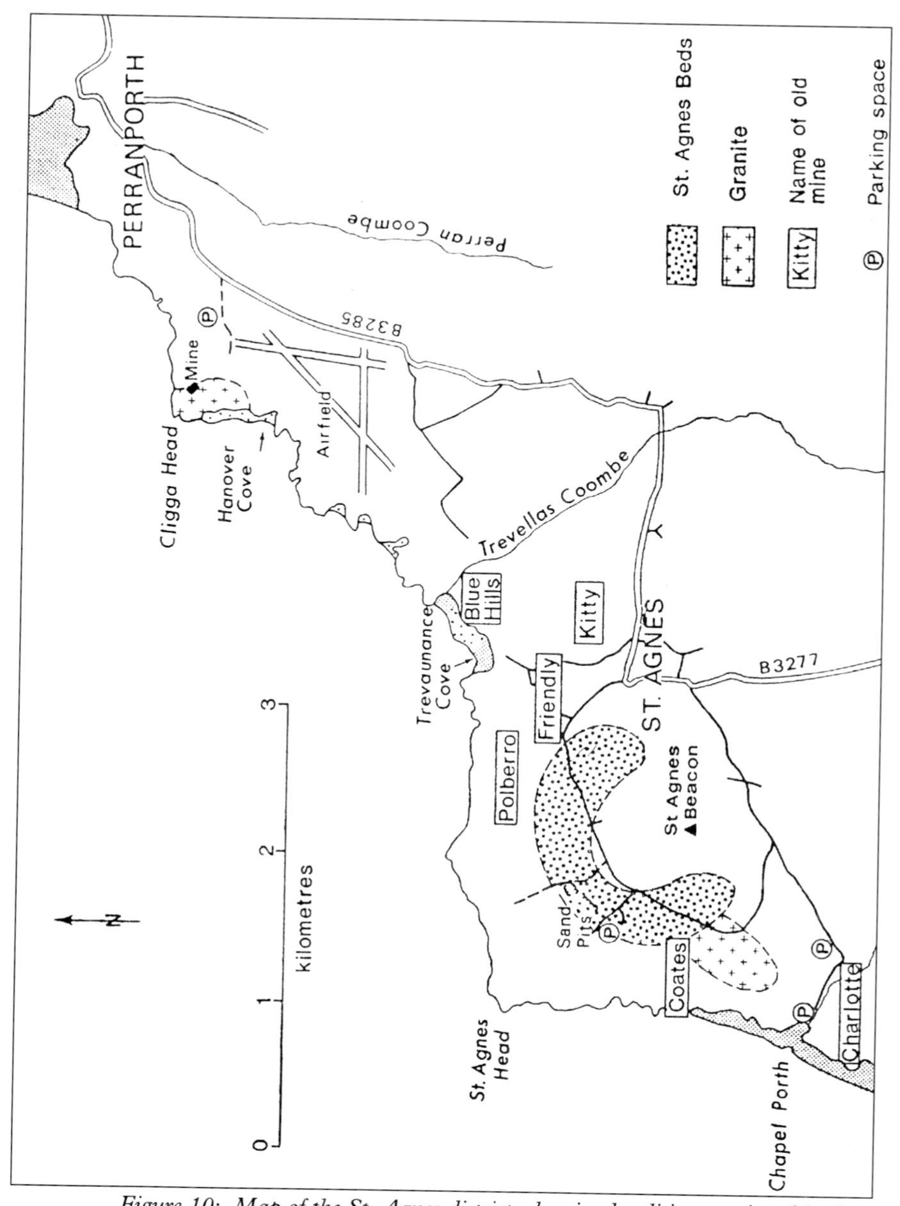

Figure 10: Map of the St. Agnes district, showing localities mentioned in the text.

Around the abandoned buildings near the 'Contact shaft' the dumps of waste material contain representatives of the veinstones and their country rocks. The granite is seen to be a coarse white porphyritic variety, grading into or veined by dark grey, finer-grained greisen (quartz-mica rock). The vein material consists principally of quartz, accompanied by chlorite and tourmaline together with ore minerals, of which cassiterite, wolframite, chalcopyrite and arsenopyrite are the most abundant.

To the northwest of the mine, a small quarry at the top of the cliffs faces northeastwards towards the distant beach of Perran Sands. In this quarry can be seen the numerous parallel greisen-bordered veins for which this locality is famous (see Back Cover). The veins themselves are mostly only a few millimetres thick, but the greisen which adjoins them ranges up to 20 cm in width, and between them the granite is lightly kaolinised. The veins dip to the north at a high angle. The geochemistry and mineralogy of the greisenisation have been described in detail by the author (Hall, 1971).

From the northern end of the quarry a path leads around the headland and down towards a small cove. About half-way down stop to observe the cliffs to the south. From here one can look along the strike of the veins above the cove. Numerous holes in the cliffs mark the entrances to old workings and to the south the cliffs are seen to be reddened by staining with iron oxides.

Agile individuals may continue down to the beach, but the route is dangerous and not recommended for parties; the path is marked by iron spikes embedded in the rock, but the lower part has fallen away. Several adits open on to the beach from the lowest level of the mine. Much quartz veinstone with platy black wolframite may be found among the granite boulders. Southwards the veins become less numerous and are not as parallel as in the quarry. The rock becomes increasingly iron-stained, and numerous patches and coatings of green alteration products draw attention to copper-bearing veins. In the cliff above, the veins can be seen to turn over into what resembles an anticlinal fold. Figure 11 shows the disposition of veins and mineralization as viewed from the west, *i.e.* the seaward side.

Return to the top of the cliff, and walk southwards to the promontory on the north side of Hanover Cove. From here the 'fold structure' of the veins can be seen in the cliffs of granite to the north. The granite and the neighbouring slates here are severely kaolinised. A small quarry above Hanover Cove shows quartz-porphyry ("elvan") underlying slates, possibly part of an upward extension of the granite. Immediately to the south, just below a boundary stone on the cliff-top

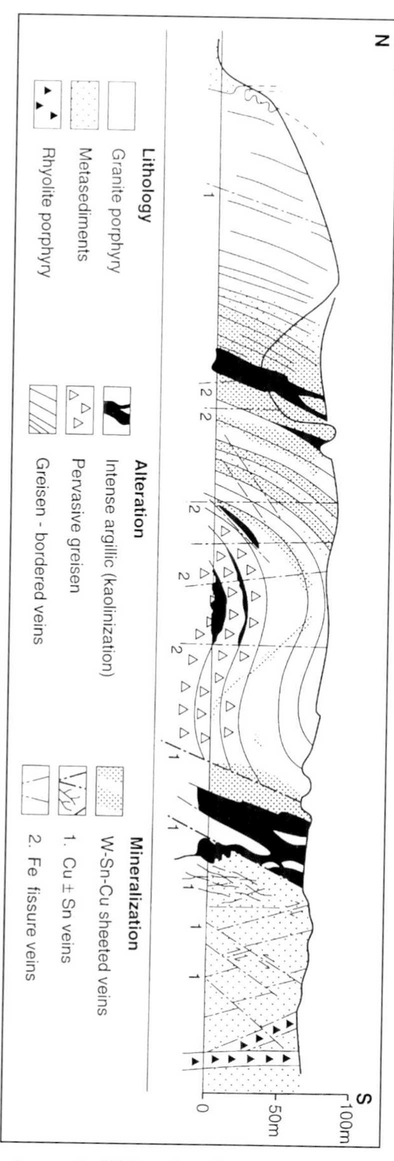

Figure 11: The north-south cliff section through the Cligga stock, showing its lithology, structure and mineralization. Adapted from Moore & Jackson (1977).

path, is a dyke of quartz-porphyry, trending east-northeast. It is about 5 m thick and dips at a high angle to the NNW. It has weathered out much more easily than the slate on either side, enabling its course to be traced across the cliffs.

Wheal Coates.

At Wheal Coates, the lodes are intersected by the coast almost at right-angles, and can be examined easily in the cliffs at Chapel Porth.

Leave St. Agnes in the direction of Porthtowan, and follow the road signposted to (St. Agnes) Beacon and Chapel Porth. If you are travelling by coach, leave it about 2 km along this road, just before a sharp right hand bend, as coaches are not permitted to use the road beyond the bend.

Walk northwards along the beach from Chapel Porth. The cliffs here are of hard dark-grey hornfels, in which more and less psammitic bands can be distinguished. The most important lode is situated directly below the engine house on the top of the cliff about 600 m north of Chapel Porth. This is the Towanwrath Lode. It has been partly eroded by the sea and partly mined away. It is surrounded by a zone of reddening. It consists of brecciated hornfels with a matrix of quartz and hematite underlying a steeply dipping quartz-porphyry dyke about 2 m thick. The hornfels and the quartz-porphyry are reddened by impregnation with hematite. The hornfelses are also locally impregnated with sulphides. Between this lode and Chapel Porth adits mark the positions of several minor lodes. The mine mainly produced tin (from cassiterite) with a small amount of copper. It is famous among mineral collectors for the perfect pseudomorphs of cassiterite after feldspar which were found here early in the last century.

From Chapel Porth return up the valley and take the left turning marked 'Beacon Drive.' About 800 m along on the left a path leads to the engine houses of Wheal Coates, which have recently been restored. An opencast working extends along a lode eastwards from the remains of a mine building 250 m NNE of the upper engine house. This lode consists of quartz veinstone with hematite, cutting tourmalinised hornfels.

St. Agnes Beacon.

Continue along Beacon Drive, and take the metalled road to the left 700 m north of Wheal Coates. Stop after 300 m and walk along a path to the right which leads past numerous shallow workings (New Downs Pits) in the St. Agnes Beds. These are unconsolidated sands and clays which lie around the north side of the hill at about the 120 m contour level. The exposures show current-bedded white sand with grey silt layers. Near the surface, the sand has weathered to a brown colour and there are dark brown bands of sand cemented by iron oxides.

The age of the St. Agnes Beds has been debated for many years. Their situation on the very flat erosion surface surrounding St. Agnes Beacon suggested the possibility of a Tertiary age, but it was not until the discovery of fossil pollen in a carbonaceous clay from the St. Agnes Formation at New Downs pit that their age was firmly established as Miocene (Walsh *et al.*, 1987). The flora indicates a Mediterranean climate with conifer forest and mixed woodland. Sedimentological features suggest that the sands are of aeolian origin.

At Beacon Cottage Farm, 1 km south of New Downs Pits, there is a deposit of clay which is not now exposed but was once worked as "candle clay". Until about 1940 this clay was sold to the mines of Cornwall for fixing candles to the walls of underground mines and to miners' hats. It has yielded an Oligocene microflora including palm pollen, suggesting a frost-free climate, and is probably a lacustrine deposit. Since the deposit is older than the St. Agnes Formation it is distinguished as the Beacon Cottage Farm Formation.

Continue along Beacon Drive, re-entering St. Agnes past the extensive dumps of Polberro tin and copper mine on the left-hand side of the road.

ITINERARY IV

FALMOUTH BAY

The coast around Falmouth is a convenient area in which to examine the Devonian sedimentary rocks in an unmetamorphosed condition. Within the Gramscatho Group, a distinction can be made between the Falmouth Beds, in which light brown sandstone predominates, and the Portscatho Beds, which are mainly grey slates. This stretch of coastline also shows two very interesting minor igneous intrusions (Figure 12). The estuary of the River Fal is a classic example of a drowned valley resulting from a rise in sea-level or subsidence of the land.

Pendennis Point.

Stop at the coastguard station on Pendennis Point (reached from the centre of Falmouth by way of Castle Drive), and walk down to Little Dennis blockhouse by the signposted path.

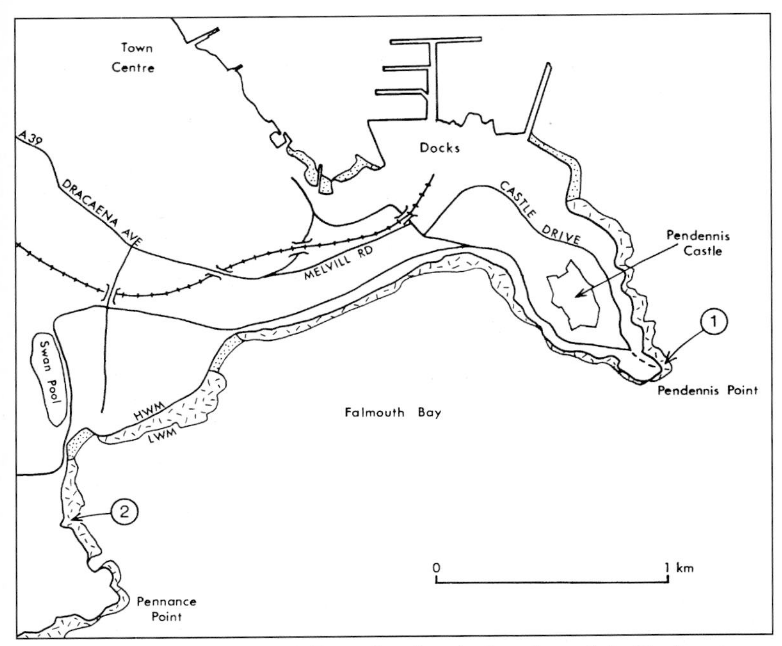

*Figure 12: Map of Falmouth Bay, showing the locations of the Pendennis minette (**1**) and the Swanpool elvan (**2**).*

Pendennis Point is formed from the Porthscatho Beds, which are grey slates with some sandstone bands, dipping at about 50° to the south. About 5 m below the blockhouse the slates are cut by a minor igneous intrusion. The rock is a minette (a mica-lamprophyre) of very unusual composition.

The minette is extremely weathered and has consequently been more strongly eroded than the surrounding slates. South of the blockhouse it takes the form of a sill about 2 m thick, but east of the blockhouse it transgresses across the bedding and becomes a dyke. It sends out several smaller offshoots into the slate. It is brown in colour and extremely rich in biotite, which can be seen in hand-specimen. Such intrusions are widely distributed in Cornwall and are believed to be hypabyssal equivalents of the Permian lavas of the Exeter Volcanic Series. The Pendennis minette is of particular petrological interest because it is very strongly peralkaline (Hall, 1982) and it contains two alkali amphiboles (riebeckite and potassium arfvedsonite) and two carbonate minerals (dolomite and ferroan magnesite), The weathering of the ferroan magnesite is responsible for the brown colour of the rock.

Swanpool.

This section extends from Swanpool beach, the most westerly of the three beaches in Falmouth Bay, southwards to Pennance Point. Pennance Point, like Pendennis Point, is formed from the relatively hard Portscatho Beds, while Falmouth Bay is eroded into the softer Falmouth Beds. At Swanpool beach the Falmouth beds outcrop on the shore. They are alternations (on a scale in the order of 1 cm to 1m) of brown sandstone and black or dark grey slate. They dip southwards at an angle of about 60° and are heavily quartz-veined. In the cliffs to the south of the beach erosion has picked out several small faults along which fault gouge is well developed.

About 300 m south of the beach, below a gully in the cliffs, the Falmouth beds are cut by a minor intrusion about 10 m across. This is a pale buff-coloured rock whose regular jointing contrasts with the contorted bedding of the sedimentary rocks. The rock is a quartz-porphyry ("elvan") dyke which has been completely greisenised. In hand-specimen crystals of quartz and muscovite and pale pink quartz-mica aggregates can be distinguished from a white saccharoidal matrix. In places the rock has broken down into a soft white sand. As long ago as 1858, Sorby calculated from his study of the fluid inclusions in quartz that the Swanpool elvan had crystallized at a depth of 53,900 feet (*c.* 15 km), a remarkable achievement for the time although probably an over-estimate.

From here onwards the rocky shore platform is backed by cliffs of brown-weathering head. Beyond a small shingle beach the Falmouth Beds give way to the hard grey Porthscatho slates. The most interesting geomorphological feature of this locality is Swanpool itself, a freshwater lake formed when the mouth of a small valley was blocked by a storm beach.

ITINERARY V

THE LIZARD PENINSULA

The Lizard peninsula is famous for the variety of igneous and metamorphic rock types which occur there, including the serpentinite which is the best example of its kind in England. Inland, the peninsula consists of a flat platform about 80 m above sea-level and there are relatively few exposures, but around the coast there are excellent sections of all the principal rock types (Figure 13). The Lizard was extensively studied by 19th century geologists and its geology was comprehensively set out in the Geological Survey maps and memoirs of 1839 (De La Beche), 1912 (Flett & Hill) and 1946 (Flett). More recent work has been summarised by Floyd *et al.*, (1993).

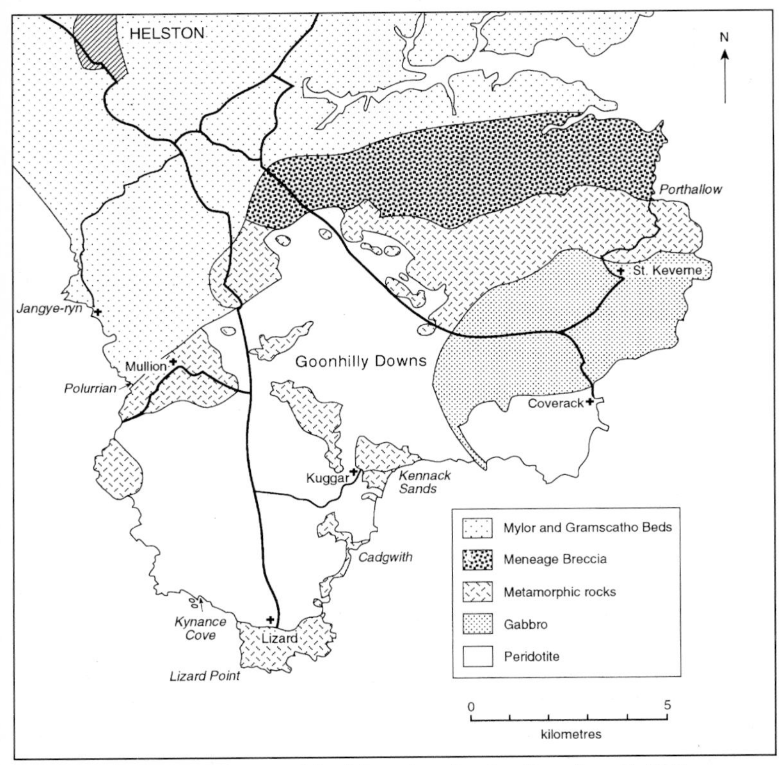

Figure 13: Geological map of the Lizard peninsula. Based on B.G.S. 1:50,000 sheet 359, by permission of the Director, B.G.S., with modifications after Smith & Leake (1984).

The principal igneous rocks of the Lizard complex are the Lizard peridotite, the gabbro and the basic dykes (black dykes). The principal metamorphic rocks are the Old Lizard Head Series (metasedimentary schists), the Landewednack hornblende schists, the Traboe hornblende schists and the Kennack gneiss. The early geologists did their best to put these rock masses into a time sequence, using such criteria as the degree of metamorphism and the intrusive relationships, but it has gradually become clear that most of the boundaries in the Lizard complex are tectonic and in recent years the geology has been fundamentally re-interpreted.

The most important new idea was the identification of the basic and ultrabasic igneous rocks as parts of an ophiolite complex, representing material originally formed on the ocean floor (Strong *et al.*, 1975). This

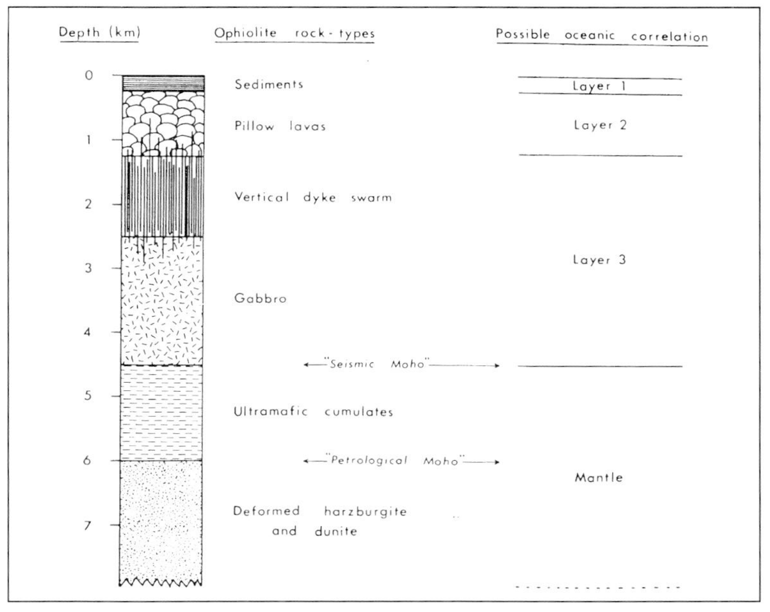

Figure 14: Idealized section of an ophiolite complex.

interpretation has been generally accepted, although there are still some difficulties fitting all the details into an ophiolite model. The major problem now is to decide how the metamorphic rocks of the Lizard are related to the ophiolitic components.

Figure 14 shows the ideal sequence of rock types present in a section through oceanic crust and upper mantle, and we can compare the Lizard assemblage with this ideal sequence. Oceanic crust is formed at oceanic spreading axes when uprising mantle peridotite undergoes partial melting to form basaltic magma. This is extruded as lava, normally pillowed because it erupts under the sea, and spilitic because it reacts with seawater. Some of the basic magma consolidates below the ocean floor as gabbro, and from the gabbro there may form ultramafic cumulates (including peridotite). A swarm of dykes is normally present at a high level in the sequence, extending partly up into the lavas, for which they are the feeders, and down into the gabbros. The lowest component of an ophiolite assemblage is deformed peridotite which represents the partial melting residue of the upper mantle. Deformation occurs as the peridotite rises towards the spreading axis and then flows laterally away from the axis.

The Lizard complex shows enough of the features of an ideal ophiolite assemblage to convince most observers that it is ophiolitic, but there are some features which are difficult to match up. Pillow lavas are present in the Lizard at various places along the northern boundary (in the Menage zone) and on Mullion Island, which is a few hundred metres off the west coast of the Lizard. Unfortunately the best exposures, on Mullion Island, are relatively inaccessible, whereas the inland exposures are not well exposed. The Mullion Island pillow lavas have chemical compositions similar to mid-ocean ridge basalts (Floyd *et al.*, 1993).

A swarm of basic dykes is present in the Lizard, but it is much less extensive than might be expected. Basic dykes with a predominant NNW-SSE orientation are present in small numbers all along the east coast of the Lizard but become abundant only north of Coverack, and only on a 1 km stretch of coast south of Porthoustock could they be described as a swarm. In this area they are so abundant as to occupy more space than their gabbroic country rock. They are certainly not as abundant as the sheeted dykes of the Troodos complex in Cyprus, which is perhaps the best known ophiolite on land, where sheeted dykes can be seen with no intervening country rock, but there are other undoubted ophiolites in which a sheeted dyke swarm is relatively poorly developed.

Gabbro outcrops over a large area in the eastern part of the Lizard peninsula and can be seen in the coastal exposures between Coverack and Porthoustock. Recent geophysical work suggests that the gabbro may not be as extensive inland as was previously thought (Floyd *et al.*, 1993), and most of its boundary is probably faulted. Some authors distinguish between the Crousa gabbro, forming the majority of the outcrop and seen on the east coast, and the Trelan gabbro seen only inland. Detailed petrographic study of the gabbro is difficult because it has been seriously affected by low grade metamorphism and hydrothermal alteration.

The most extensive of the igneous rocks is the Lizard peridotite. This may be seen at many localities around the coast and has been quarried inland. It is always partly or completely serpentinised and is usually black due to the presence of secondary magnetite disseminated through the serpentinised rock. Reddish or greenish colours are seen in particularly weathered serpentinite and these varieties are used in the manufacture of "serpentine" ornaments. In petrological terms the peridotite varies in composition (disregarding serpentinisation) from dunite (olivine only) through harzburgite (olivine + orthopyroxene) to lherzolite (olivine + orthopyroxene + clinopyroxene) and small amounts of amphibole and chromite are often present. In hand specimen one can make a distinction between two types of peridotite: (1) the coarse-grained "bastite-serpentinite", which contains large shiny crystals of orthopyroxene (commonly pseudomorphed by serpentine) in a fine-grained groundmass and (2) fine-grained varieties (the "tremolite-serpentinite" and "dunite-serpentinite" of early authors). According to Green (1964a, b), the bastite-peridotite represents a primary mineral assemblage which formed under upper mantle conditions, and the fine-grained peridotite is the product of recrystallization at lower pressures. The serpentinization is generally agreed to be a late-stage, low-temperature type of alteration which may even still be going on.

In terms of the ophiolite model (Figure 14) it is not clear how much of the Lizard peridotite should be regarded as an ultramafic cumulate and how much as a deformed melting residue. For the most part it is strongly foliated and there is little evidence of cumulus textures (Rothstein, 1988). On the other hand, there are some basic layers within the peridotite near its northern margin (Leake & Styles, 1984), and there are alternations of basic and ultrabasic layers in the Traboe hornblende schists at Porthkerris which could be interpreted as metamorphosed cumulates.

Structural studies, and evidence from Geological Survey boreholes, have shown that the Lizard peridotite has a sheet-like form, having been thrust over the metamorphic rocks which occur along the southern and southeastern coastline of the peninsula.

The metamorphic rocks of the Lizard are divided into four groups: (a) the Old Lizard Head Series — metasediments of varied composition, (b) the Landewednack hornblende schists — originally basic lavas or tuffs with some interbedded sediments, now metamorphosed to amphibolites, (c) the Traboe hornblende schists — differing from the Landewednack schists in relative scarcity of epidote, and (d) the Kennack gneiss.

The Old Lizard Head Series are best seen at Lizard Point and are predominantly mica-schists with some interbedded quartzites and hornblende schists.

The Landewednack schists are hornblende-schists or hornblende-gneisses, often rich in epidote. They have the chemical composition of basalts and are presumed to be metamorphosed basaltic lavas.

The Kennack Gneiss is typically a banded rock in which there are dark bands of hornblende schist very similar to the Landewednack schists together with light bands of pink quartzo-feldspathic gneiss. The proportion of light and dark bands varies; in some inland outcrops the quartzo-feldspathic gneiss predominates, but often the two components are present in roughly equal amounts. The origin of the Kennack Gneiss has been much debated. Two alternative interpretations have received wide support: the first is that the banding results from the flow of a mixture of acid and basic magmas (e.g. Flett, 1946); the second is that the gneisses were produced by the metamorphism and anatexis of a mixture of sediments and basaltic rocks comparable to the Old Lizard Head Series and Landewednack hornblende schists, the heat having been supplied by the emplacement of the hot peridotite (e.g. Malpas & Langdon, 1987).

The Traboe hornblende schists are the most problematic group of rocks in the Lizard complex. They differ from the Landewednack hornblende schists in being generally coarser, having a steeply dipping foliation, and in not containing epidote. Although essentially meta-basic, they are more variable in chemical composition than the Landewednack schists. They occur at various places in the northern part of the Lizard complex, for example at Porthkerris. There are two main interpretations of the Traboe schists. Green (1946b, c) regarded them as the product of contact metamorphism of the Landewednack schists by the peridotite. Later workers have regarded them as metamorphosed gabbroic rocks forming part of the ophiolite complex.

Not all of the main rock types of the lizard complex have been dated isotopically, and most are represented only by K-Ar dates which may have been affected by resetting, but the two following ages (both Devonian) go some way to clarifying the development of the complex:-

Olivine gabbro, Coverack	375±34 m.y. Sm-Nd	Davies (1984)
Kennack Gneiss (acid vein)	369±12 m.y. Rb-Sr (WR)	Styles & Rundle (1984)

Assuming the ophiolite interpretation of the basic and ultrabasic rocks to be correct, the gabbro date represents the time of formation of the original oceanic crust, and if the Kennack gneiss is the result of melting associated with peridotite emplacement, the implication is that this segment of oceanic crust was obducted onto continental crust very soon after it formed.

North of the faults which bound the Lizard complex, the igneous and metamorphic rocks are separated from normal Devonian sediments by a belt of rock known as the Meneage Breccia. This formation consists of slates, breccias and included masses of volcanic rock, hornblende schist, quartzite and various other materials, sometimes as large as hundreds of metres across, among which can be recognised both Devonian and pre-Devonian constituents. The Meneage Breccia has been referred to as the ‘Meneage crust zone’, implying a tectonic origin, but it is now considered more likely that it is a sedimentary melange. The rocks of the Lizard complex are generally believed to have been thrust over the sediments to the north, but where the actual boundary is seen (at Polurrian Cove) it is a simple reverse fault.

All the best exposures of the Lizard peninsula are on the coast, and the itinerary takes the form of a coastal tour from west to east.

Jangye Ryn.

Jangye Ryn (658207) is reached from the Helston-Lizard road by taking a side road signposted to Gunwalloe. The road ends at Gunwalloe Church, and Jangye Ryn is the rocky cove on the *north* side of the church. Some larger coaches would find the road too narrow, in which case the locality is easily reached by the coastal path from Poldhu.

The section shows alternating beds of slate and greywacke belonging to the Gramscatho Beds. Near the northern end of the beach are beautiful examples of large angular folds overturned to the northwest. These structures made a great impression on the earliest geologists to visit the locality, who were unable to comprehend their origin. Sedgwick, in 1822, described them as being ". . difficult to account for by the action of mere mechanical forces." In addition to the folding there is also a large fault shown by brecciation and by quartz-veining in the neighbouring sediments. Among the quartz veins can be found good examples of S-shaped quartz-filled tension gashes.

Fragments of fossil plant material can be seen in some of the greywackes and for many years they provided the only evidence that the age of these beds was Devonian. They are now known from micropaleontological evidence to be Upper Devonian (Le Gall *et al.*, 1985).

Polurrian Cove.

On the west side of the peninsula the junction between metamorphic rocks of the Lizard complex and the Devonian sediments to the north is a fault, which reaches the sea at Polurrian Cove. There is nowhere to park at this locality and it is best approached by walking from Mullion Cove, or if travelling by coach your driver can leave you at Polurrian Hotel and pick you up at Mullion Cove.

The north side of the cove shows dark blue slates of the Gramscatho Beds, similar to those at Jangye Ryn. On the south side of the cove are dark blue or green feldspathic hornblende schists of the Traboe type. The two are clearly separated by a zone of fault breccia dipping at about 50° to the southeast. The hornblende schist lies above the slate and the fault is therefore a reverse one. The fault is about 60 m south of where the stream reaches the beach.

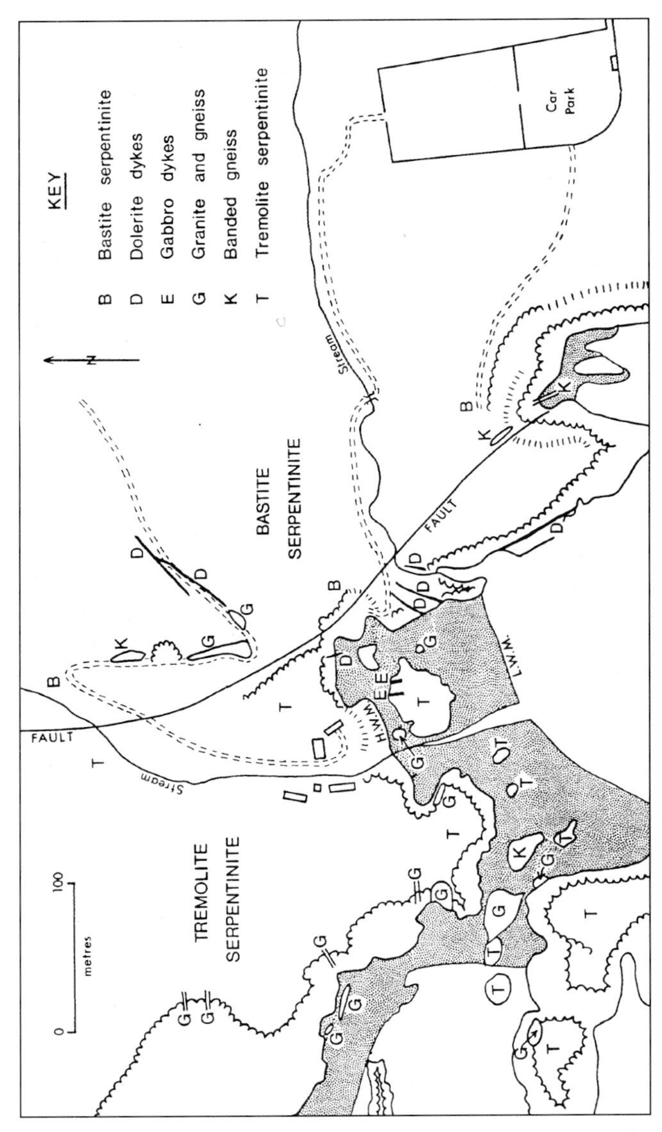

Figure 15: Map of Kynance Cove, showing details of the exposures. The country rock is serpentinite, but there are numerous veins and small masses of granite, epidiorite and dark banded gneiss. Modified from Flett (1946, fig. 7).

Kynance Cove.

At Kynance Cove you can see the two main varieties of the Lizard peridotite. The primary type (bastite serpentinite) occurs on the east side of the cove. The recrystallised type (tremolite serpentinite) outcrops on the beach and on the western side of the cove (Back Cover, top). The two are separated by a NNW-SSE fault, marked by brecciation, and both varieties are cut by small bodies of granite, gneiss and epidiorite. The tremolite serpentinite is a fine-grained banded rock, whereas the bastite serpentinite is coarse-grained with large shiny platy crystals of enstatite (bastite) which give the rock a pitted appearance on weathered surfaces. Details of the exposures are shown in Figure 15.

Lizard Point.

The coastline to the east of Lizard Point shows exposures of the various metamorphic rocks which were previously thought to pre-date the serpentinite intrusion and are now thought to underlie the serpentinite. The shore can be reached by taking a path which leads from the 'most southerly point in England' (signposted) to the old lifeboat station in Polpeor Cove.

The Old Lizard Head Series which are seen at Polpeor Cove are folded muscovite-, chlorite-, epidote- and hornblende-schists, together with impure quartzites. Bedding is well marked, although it is believed that some of the green schists may originally have been volcanic ashes and some of the more massive hornblende schists could have been lava flows or sills.

There are small car parks at the 'most southerly point' and at the lighthouse, but coaches should be left at Lizard town when visiting either Lizard Point or Landewednack.

Landewednack.

Park at Landewednack Church (Figure 16), and walk down to Church Cove (715127), in which are exposed hornblende schists of the Landewednack type. These rocks are distinguished from the Traboe hornblende schist seen at Polurrian by the presence of greenish-yellow epidote. In composition, the hornblende schists are basaltic, but is not known whether they were originally volcanic rocks contemporaneous with the Old Lizard Head Series or whether they were intruded as sills into the latter.

Inland from the cove, 100 m north of the stream, there is a small quarry next to the coastal path. This leads into a larger quarry facing the sea, showing bastite serpentinite cut by talc veins. On the south side of the larger quarry can be seen the junction between the serpentinite

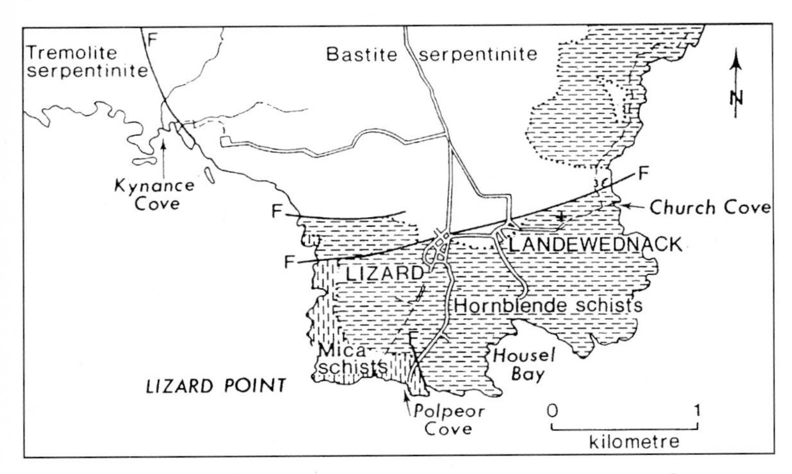

Figure 16: Geological map of the area around Lizard Point. Based on Sanders (1955).

and hornblende schist. Both rock types are reddened, brecciated, and veined with calcite and dolomite near the contact, which is a composite fault zone dipping to the north. The serpentinite in the northeast corner of the quarry is cut by sheets of a red banded gneiss.

In Parn Voose Cove, 150 m to the north, the serpentinite can be seen to overlie gneisses and migmatites, which are well exposed at sea-level. The gneisses on the north side of the bay are banded, with two discrete components: (a) a pink quartzo-feldspathic rock and (b) a black amphibolitic rock. The gneisses at the southern end are more irregularly banded mica-bearing gneisses, looking rather more like metasediments. On the northern side there are also sheets of pink granitic rock and irregular bodies of gabbro, often very coarse grained and sometimes foliated. In places the gabbro is cut by acid veins from the banded gneiss.

Cadgwith.

Cadgwith is a tiny village at the foot of a steep valley and it is best approached via Chyheira (709143). There is a car park above the village (718148), beyond which vehicles risk getting stuck.

Cadgwith Cove shows excellent exposures of the Landewednack hornblende schists. They are banded amphibolites with conspicuous pale green epidote-rich bands. The high ground above the village is on serpentinite, which therefore appears to overlie the Landewednack schists.

From the small harbour of Cadgwith follow the coastal footpath for 600 metres to the northeast and descend towards the shore at Enys Head (728149). This is probably the best locality in the Lizard for studying rodingites. These are calc-silicate veinstones and altered rocks which occur in all serpentinites in small amounts and are also present in the modern oceanic crust. The most conspicuous example at Enys Head takes the form of white veins from 5 to 30 cm thick cutting the serpentinite. Under the microscope these prove to be almost entirely made up of garnet (hydrogrossular), with minor amounts of serpentine, ilmenite and sphene (Hall & Ahmed, 1984). Also present in the serpentinite at this locality are very dark coloured basic dykes which have not only been metamorphosed, but also metasomatised, so that their mineralogical composition now consists of varying proportions of amphibole, serpentine, chlorite and prehnite. Both the garnetite veins and the metasomatised basic dykes are the result of hydrothermal activity, which was probably associated with the process of serpentinization and could possibly have occurred while the enclosing serpentinite still formed part of the ocean floor.

Kennack Sands.

The best exposures of the Kennack gneiss are at Kennack Sands. The gneiss is variable in composition, but usually banded, with a dark basic component and a light quartzo-feldspathic acid component, the two sometimes being intimately mixed to give a relatively homogeneous rock of intermediate composition. In addition to occurring as bands within the banded gneiss, there is also some acid material which cuts across the banding of the banded gneisses and also forms veins within the peridotite.

At the west side of Kennack Sands is a well-known exposure (Figure 17) which shows the relationships between the gneiss and the other major rocks of the Lizard complex. The exposure is on the edge of the beach, immediately south of the place where the road reaches the shore (734165). Here, the cliff is made of bastite serpentinite, part of a large mass about a hundred metres in length which is surrounded by banded gneiss. Veins of gneiss penetrate into the serpentinite, and blocks of the latter are enclosed by the gneiss. The banding of the gneiss is parallel to the contact with the serpentinite, but the foliation in the latter is truncated by the gneiss. The serpentinite is cut by a gabbro dyke, which is itself stepped by a small fault and cut by a basalt dyke, which in turn is truncated by the gneiss. The sequence thus appears to be: serpentinite - gabbro - basalt - gneiss. The relationship between the gneiss and the serpentinite is not, however, a simple intrusive one. It is difficult to see how such a composite rock as the gneiss could have intruded magmatically into the serpentinite. It is more likely that the serpentinite bodies at Kennack have been tectonically included in the gneiss. Another feature of this outcrop which is difficult to explain is that the basalt (epidiorite) dykes in the serpentinite are petrographically indistinguishable from the dark component of the surrounding banded gneiss.

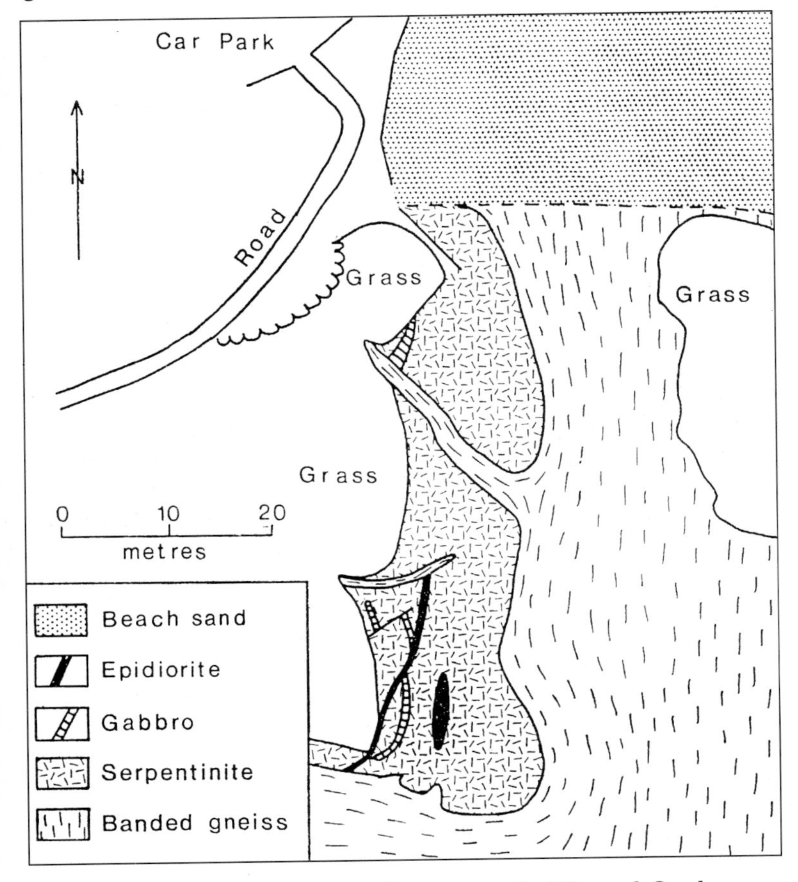

Figure 17: Map of the exposures at the western end of Kennack Sands immediately south of the point at which the road reaches the shore. Although there is a car park here, the road beyond Kuggar (724163) is too narrow for coaches. Modified from Flett (1946, fig. 15).

The serpentinite has been somewhat altered near the contact with the surrounding acid material, and veins of talc and asbestos are relatively well developed at this locality. About 100 m SW of the end of the road anthophyllite-asbestos can be found in fibres up to several centimetres in length.

Coverack.

The shore at Coverack is of outstanding interest for the relationships between gabbroic and ultrabasic rocks and for their interpretation as ophiolite components. However, it takes a little time to distinguish all the rock types, especially where they are partly covered by seaweed. You may find it easier to identify the various rocks if you start at the northern end of the section and work southwards towards the harbour.

The shore northwards from the main car park (where the main road reaches the coast) shows gabbro with occasional peridotite xenoliths. The gabbro is most easy to recognise when it is slightly altered, because the feldspar in the fresh gabbro has such a dark colour. The rock is composed of plagioclase and augite with minor olivine and ilmenite. Many of the boulders of gabbro on the shore show a structure described by Flett (1946) as "brecciform". It appears to consist of blocks of a dark coloured gabbro set in a matrix of lighter coloured gabbro, leading Flett to the view that the gabbro had been fractured while still hot and then bound together by further injection of a similar but more leucocratic magma. In fact, the lighter matrix of the brecciform gabbro does not represent a second injection of magma, but is due to postmagmatic alteration concentrated about a number of interlacing planes. This feature was not immediately recognised as a type of alteration because the zones of alteration do not correspond to the present directions of jointing, and only rarely does the alteration product appear in discrete veins. The lighter coloured gabbro has been partially replaced by prehnite, and the darker coloured gabbro is simply its unaltered equivalent. The white material in the most altered examples is composed almost entirely of fine-grained colourless prehnite.

Southwards from the car park the predominant rock type exposed on the shore is peridotite. The rock is the type described by Flett (1946) as "bastite-serpentine", with conspicuous large crystals of orthopyroxene. The rock is a true peridotite, having been less severely serpentinized here than in other parts of the Lizard complex. The peridotite is a dark, mainly greenish rock but reddened where it is particularly weathered. It is cut by numerous sheets of gabbro and pegmatitic gabbro and by basalt dykes, and it contains many small bodies of dunite which are reddish and lack orthopyroxene. Some of the gabbroic sheets are severely sheared or even mylonitized.

Towards the southern end of the section, within 200 m of the harbour, occurs the troctolite for which Coverack is a classic locality. The rock is composed of varying proportions of red serpentinized olivine and white plagioclase giving it a distinctive appearance. The name troctolite means 'trout-stone', indicating its supposed resemblance to the skin of a trout. The troctolite was mapped by Flett (1946) as irregular masses or small bosses, and by Floyd *et al.* (1993) as sheet-like bodies. Like the peridotite, the troctolite is cut by gabbro sheets and basalt dykes.

The status of the troctolite is crucial for petrological interpretation. The composition of the troctolite does not correspond to any present-day volcanic rock, suggesting that there is no magma of this composition, although there are some layered basic intrusions whose calculated parent magma is troctolitic (Morse, 1981). On the other hand, troctolites do occur as cumulate layers in layered basic intrusions (e.g. Rhum, Stillwater). A problem with interpreting the Coverack troctolite as a cumulate is that some troctolite veins apparently cut the peridotite, and there are peridotite xenoliths in the troctolite. Some of the well-known exposures at the base of the cliff which previously showed these relationships have been covered by extensions of the sea-wall, but others can still be seen on the shore.

Porthoustock.

The coast on the south side of Porthoustock cove shows the greatest concentration of basic dykes in the Lizard complex. The cove itself conceals the faulted contact between hornblende schist to the north and gabbro to the south. The dyke swarm is seen more clearly on Porthoustock Point than in the nearby large quarry.

On Porthoustock Point, fine-grained basic dykes aligned NNW-SSE cut coarse-grained gabbro. The proportion of dykes to host rocks is approximately 50:50, but the abundance of dykes falls rapidly south of Porthoustock and this is the only place where they can be described as a dyke swarm. The dykes vary from a few cm to more than a metre in width, but cannot be followed for any distance because of the amount of shearing which has taken place. The chemical composition of the dykes is basaltic, but although many of them look unaltered they have a low grade metamorphic mineralogy. Chilled margins are not conspicuous, probably because of the metamorphism and perhaps also because of the shearing. The host gabbro has undergone considerable hydrothermal alteration and is cut by calcite veins; these features can be examined in the large blocks adjoining the roadway in the upper quarry.

Ascend the coastal path for 300 metres northeastwards from Porthoustock to the first of several large quarries in hornblende schists of Landewednack type. The rock here is banded and very dark in colour, with a gently-dipping foliation. There are no basic dykes at all, in contrast to their abundance on the south side of the cove.

Porthallow.

Porthallow Cove is on the northern boundary of the Lizard complex. The section on the north side of the cove extends for 2 km northwards through the Meneage Breccia, which is a very unusual and puzzling formation. It consists predominantly of a dark, well-cleaved breccia containing fragments of slate and other rock types, together with larger masses up to hundred of metres across of various sedimentary, igneous and metamorphic rocks, all of types which occur in the Lizard area. In the past, the breccia was considered to be tectonic in origin, but a sedimentary origin for the formation now seems much more likely.

On the north side of Porthallow beach the breccia appears very similar to many other Devonian slates in this part of Cornwall, but farther along the shore many other distinctive types of rock are found. In the cliff on the north side of Porthallow Cove there is a mass of grey igneous rock (sodic microgranite) about 30 metres wide which reaches to the top of the cliff. In Nelly's Cove, immediately north of Porthallow Cove, there occur pillow lava and bedded radiolarian chert. At Fletching's Cove, about 800 m north of Porthallow, there is a limestone which was at one time believed to be Silurian from its fossil content, although more recent examination of the fossils casts doubt on this identification. At numerous places along the section there are examples of the Meneage quartzite, a hard white rock which is easily recognised among the other constituents of the breccia. It sometimes occurs in very large masses, especially north of Fletching's Cove, and blocks of this rock are abundant on the shore where they have been washed out of the breccia. There are few fossils in this quartzite, but those that have been found indicate an Ordovician age.

On the east side of Porthallow Cove the exposures are very difficult to interpret. The boundary of the Lizard complex is described on the Geological Survey map as a definite fault, and there is certainly plenty of shearing here, but the actual fault plane cannot easily be located. Slates very similar to those on the north side of the beach occur in the rocks east of the beach, and can be assigned to the Meneage Breccia. In a distance of 200 m eastwards from the beach the shore and cliff exposures show slate, phyllite, mica schist, hornblende schist, fine-grained green serpentinite, and pink acid gneiss, none of whose mutual relationships are easy to determine. The exact positioning of the Lizard boundary inland from the shore depends partly on the identification of the phyllites, which have alternatively been interpreted as slates belonging to the Meneage Breccia or as mica schists belonging to the Lizard complex.

Porthkerris.

The complicated mixture of rock types seen at Porthallow is succeeded eastwards by "hornblende schists", which have been

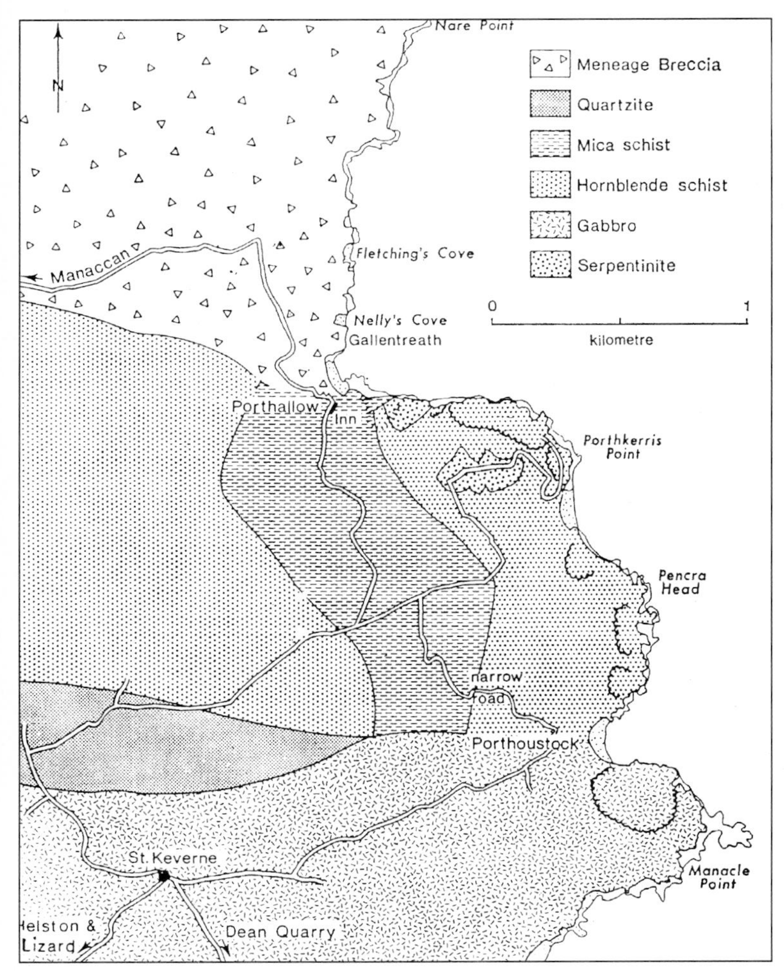

Figure 18: Map of the northeastern margin of the Lizard complex.

extensively quarried at Porthkerris Point (805230). The quarries can be reached by walking from Porthallow, or by road along a narrow turning off the Porthallow-Porthoustock road (Figure 18). They contain a great variety of coarse and fine, light and dark amphibolites and mafic granulites. Some of the joint surfaces are densely encrusted with small but perfect crystals of adularia, partly covered by calcite.

The distinction between the Landewednack and Traboe hornblende schists can be seen around Porthkerris Point. On the south side of Porthkerris Cove (806227), the hornblende schists are of the Landewednack type, *i.e.* fine-grained, epidote-bearing, and with a foliation dipping uniformly to the NW at about 30°. In the quarries north of Porthkerris Point the schists are of the Traboe type, *i.e.* coarse-grained, with a steeply dipping and folded foliation. They are often banded, and contain some layers of ultrabasic composition, *i.e.* peridotite or pyroxenite.

REFERENCES

CHEN, Y., CLARK, A.H., FARRAR, E., WASTENEYS, H.A.H.P., HODGSON, M.J. & BROMLEY, A.V. 1993. Diachronous and independent histories of plutonism and mineralization in the Cornubian batholith, southwest England. *J. Geol. Soc. London,* **150,** 1183-1191.

CHESLEY, J.T., HALLIDAY, A.N., SNEE, L.W., MEZGER, K., SHEPHERD, T.J. & SCRIVENER, R.C. 1993. Thermochronology of the Cornubian batholith in southwest England: implications for pluton emplacement and protracted hydrothermal mineralization. *Geochimica et Cosmochimica Acta,* **57,** 1817-1835.

DARBYSHIRE, D.P.F. & SHEPHERD, T.J. 1985. Chronology of granite magmatism and associated mineralization, S. W. England. *J. Geol. Soc. London,* **142,** 1159-1177.

______& ______1987 Chronology of magmatism in south-west England: the minor intrusions. *Proc. Ussher Soc.,* **6,** 431-438.

DAVIES, G.R. 1984. Isotopic evolution of the Lizard Complex. *J. Geol. Soc. London,* **141,** 3-14.

DINES, H.G. 1956. *The Metalliferous Mining Region of South-West England.* Mem. Geol. Surv. G. B., 2 vols.

EDMONDS, E.A., McKEOWN, M.C. & WILLIAMS, M. 1969. *South-west England* (3rd edition). British Regional Geology Handbook.

EMBREY, P.G. & SYMES, R.F. 1987. *Minerals of Cornwall and Devon.* British Museum (Natural History), London.

FLETT, J.S. 1946. *Geology of the Lizard and Meneage* (explanation of sheet 359). 2nd edition. Mem. Geol. Surv. U.K.

FLOYD, P.A., EXLEY, C.S. & STYLES, M.T. 1993. *Igneous rocks of South-West England.* Chapman and Hall, London.

GREEN, D.H. 1964a. The petrogenesis of the high-temperature peridotite intrusion in the Lizard area, Cornwall. *J. Petrology,* **5,** 134-88.

______1964b. A re-study and re-interpretation of the geology of the Lizard Peninsula, Cornwall. Pp. 87-114 in: *Present Views on Some Aspects of the Geology of Cornwall and Devon* (ed. K. F. G. Hosking and G. J. Shrimpton). Royal Geol. Soc. Cornwall.

______1964c. The metamorphic aureole of the peridotite at the Lizard, Cornwall. *J. Geol.*, **72**, 543-63.

HALL, A. 1971. Greisenisation in the granite of Cligga Head, Cornwall. *Proc. Geol. Assoc.* **82**, 209-230.

______1982. The Pendennis peralkaline minette. *Mineralogical Magazine*, **45**, 257-266.

______ & AHMED, Z. 1984. Rare earth content and origin of rodingites. *Chem. Erde*, **43**, 45-56.

______ & JACKSON, N.J. 1975. Summer field meeting in West Cornwall, 15-20 September 1974. *Proc. Geol. Assoc.*, **86**, 95-102.

HOSKING, K.F.G. 1957. The vein system of St. Michael's Mount, Cornwall. *Trans. Royal Geol. Soc. Cornwall*, **18**, 493-509.

LEAKE, R.C. & STYLES, M.T. 1984. Borehole sections through the Traboe hornblende schists, a cumulate complex overlying the Lizard peridotite. *J. Geol. Soc. London*, **141**, 41-52.

LE GALL, B., LE HERISSE, A. & DEUNFF, J. 1985. New palynological data from the Gramscatho Group at the Lizard front (Cornwall): palaeogeographical and geodynamical implications. *Proc. Geol. Assoc.*, **96**, 237-253.

LEVERIDGE, B.E. & HOLDER, M.T. 1985. Olistostromic breccias at the Mylor/Gramscatho boundary, south Cornwall. *Proc. Ussher Soc.*, **6**, 147-154.

MALPAS, J. & LANGDON, G.S. 1987. The Kennack gneisses of the Lizard complex, Cornwall, England: partial melts produced during ophiolite emplacement. *Canadian J. Earth Sci.*, **24**, 1966-1974.

MOORE, J. McM. & JACKSON, N. 1977, Structure and mineralization in the Cligga granite stock, Cornwall. *J. Geol. Soc. London*, **133**, 467-480.

MORSE, S.A. 1981. Kiglapait geochemistry IV: The major elements. *Geochimica et Cosmochimica Acta*, **45**, 461-479.

ROTHSTEIN, A.T.V. 1988. An analysis of the textures within the primary assemblage peridotite, the Lizard, Cornwall. *Proc. Geol. Assoc.*, **99**, 181-192.

SANDERS, L.D. 1955. Structural observations on the South-East Lizard. *Geol. Mag.*, **92**, 231-40.

SMITH, K. & LEAKE, R.C. 1984. Geochemical soil surveys as an aid to mapping and interpretation of the Lizard Complex. *J. Geol. Soc. London,* **141,** 71-78.

SORBY, H.C. 1858. On the microscopical structure of crystals, indicating the origin of minerals and rocks. *Quart. J. Geol. Soc.,* **14,** 453-500.

STONE, M. 1968. A Study of the Praa Sands elvan and its bearing on the origin of elvans. *Proc. Ussher Soc.,* **2,** 37-42.

______1969. Nature and origin of banding in the granitic sheets of Tremearne, Porthleven, Cornwall. *Geol. Mag.* **106,** 142-158.

______1975. Structure and petrology of the Tregonning-Godolphin granite, Cornwall. *Proc. Geol. Assoc.,* **86,** 155-170.

STRONG, D.F., STEVENS, R.K., MALPAS, J. & BADHAM, J.P.N. 1975. A new tale for the Lizard (abstract). *Proc. Ussher Soc.,* **3,** 252.

STYLES, M.T. & RUNDLE, C.C. 1984. The Rb-Sr isochron age of the Kennack Gneiss and its bearing on the age of the Lizard Complex, Cornwall. *J. Geol. Soc. London,* **141,** 15-19.

TAYLOR, R.T. & GOODE, A.J.J. 1987. Late Pleistocene and Holocene radiocarbon dates from the Penzance district, Cornwall. *Proc. Ussher Soc.,* **6,** 559.

TILLEY, C.E. 1935. Metasomatism associated with the greenstone-hornfelses of Kenidjack and Botallack, Cornwall. *Mineralogical Magazine,* **24,** 181-202.

TURNER, R.E. TAYLOR, R.T., GOODE, A.J.J. & OWENS, B. 1979. Palynological evidence for the age of the Mylor Slates, Mount Wellington, Cornwall. *Proc. Ussher Soc.,* **4,** 274-283.

WALSH, P.T., ATKINSON, K., BOULTER, M.C. & SHAKESBY, R.A. 1987. The Oligocene and Miocene outliers of west Cornwall and their bearing on the geomorphological evolution of Oldland Britain. *Phil. Trans. Royal Soc.,* **A, 323,** 211-245.